KW-482-731

CONTENTS

What Price Abortion ?

BY

ROSCOE HOWELLS

GOMERIAN PRESS

1973

First Impression – January 1973

SBN 85088 184 6

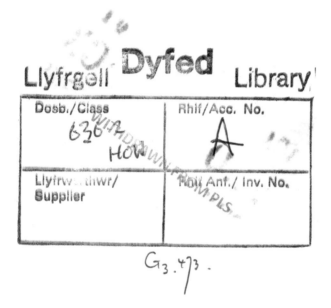
PRINTED BY J. D. LEWIS AND SONS LTD.
GWASG GOMER, LLANDYSUL

Chapter I

SWEEPING IT UNDER THE CARPET

WHEN the headlines proclaim yet another story about abortion, the reaction is likely to be anything from righteous indignation or outright condemnation to a leering desire for the lurid details. Not often, I suspect, is a great deal of thought given to whatever human misery might be behind the individual cases. Such abortions, of course, are in humans and will have been induced for what some people will consider to have been sufficiently good reasons.

What is not often realised is that abortion in cattle can also affect humans and in more ways than one lead to a whole world of human misery. Even now, when there has been considerable talk about it under its proper name, brucellosis, far too little is being done about it and the time has long since passed when more people in authority should be paying much more attention to what is involved. And there is need for it to be done because it concerns people. It is an old-fashioned and out-moded idea, I know, but I am firmly convinced that, even in this computerised, super-efficient age, people matter. And, although brucellosis is thought of chiefly in terms of a cattle disease, it is becoming of greater importance to people in its effects upon them with almost every passing day.

It is only of comparatively recent years that realisation has dawned that it can affect humans, and the hidden danger lies partially in the fact that so little is known about it. At a guess I would say that the medical profession knows rather less about the disease in humans than the veterinary profession knows about it in cattle. And that is not a great deal. To say this is not in any way to denigrate either body of people but merely to emphasise, not only the lack of research, but the fact that funds have never been made available for it. Nor is it to say that they do not know anything at all. But, compared with what is known about so many other diseases, it is little enough. Indeed it is surprising what divergent views are held by vets on the subject, and I know this because I have had the opportunity of talking to so many of them. Likewise the doctors also vary considerably in what they know, claim to know or admit to not knowing, about it in humans. It is, some say, very much in the mind and the imagination and a bit of a fashionable fad. Like it was about twenty years ago, the fashionable thing to talk about slipped discs whenever anybody had something wrong with them which could not be diagnosed.

For years I had suffered at odd times from rheumatism, fibrositis, or call it what you will, in my shoulder and arm. Eventually a Ministry of Agriculture vet told me he thought it was a slipped disc. At the time, our doctor was on holiday so I put the proposition to his locum.

"Ah, yes, of course," he smiled, "that's a very fashionable idea, but I'm afraid there are not nearly as many cases about as some people fondly imagine. So I'll give you a bottle of medicine and we'll see how it gets on."

I took his medicine, of course, because I'm like that, but presumably he had given me the proverbial bottle of coloured water which he would have prescribed for any other hypochondriac. It certainly didn't do any good. So when our own doctor came back, and fortunately he was a man with considerable experience in that particular field, I put the Ministry vet's proposition to him and he told me to move my neck this way and that, and in no time at all he said that this particular vet was obviously no fool and knew what he was talking about. When the specialist saw me he held up the X-ray plates, smiled sadly and said, "Ah, yes. Another rugby player. Occupational hazard I'm afraid. All this tackling."

The first part of the treatment was to have a collar round my neck, which for six weeks made me look like some sort of extraordinary curate walking about the place, and you'd never believe how many fools and slow-bellies there are in the world who seem to find great amusement in the oddest things, particularly someone else's misfortunes. When they had finished with the treatment of hot lamps and pulling and prodding they said, "There you are then, that's the best we can do for you. Now go and learn to live with it." But they said nothing about it being all in the mind.

So when I hear anyone say that all this talk about brucellosis is just a bit of a fashionable fad I am not unduly impressed, because, if what I've had is just a fashionable fad, then anybody is welcome to it.

The bug, Brucella Abortus, to give it its proper name, causes the disease brucellosis in cattle. The effect is for a cow to abort or calve prematurely, hence it has for years been known to the farming community as contagious abortion. All abortions, of course, are not contagious. They can be caused through a variety of reasons and do not usually present any real problem. But when it is contagious and affects a whole herd, and over a long period, the results are disastrous.

The main value to the farmer of the dairy cow is for her milk and she gives the greatest quantity of this when she is newly calved. If she aborts, however, she does not give as much milk and this is the first loss. Invariably the abortion occurs at such a stage in pregnancy that the calf is dead and this is the second loss. If the farmer is looking for replacements for his dairy herd he needs the heifer calves to rear. On the other hand, much of the beef produced in this country today comes from the dairy herd and many farmers use beef bulls on their cows to produce calves for this purpose. Either way, therefore, the loss of the calf is serious. Far worse than this, however, and this is something which is not fully appreciated by those who have not experienced the trouble, is a much greater loss than the first two put together. One of the side effects of brucellosis in cattle is that it often causes the cow to become infertile. Treatment of this can be costly and often ineffective so that the farmer, having thrown good money after bad, finds himself with a cow on his hands giving no milk. At this stage he is faced with the realisation that even if he does manage to get her in-calf it will be at least nine months before she can calve and again come into milk, and she will have cost a great deal to feed whilst producing nothing. Added to this is the knowledge that if she does calve normally she will, every time she calves, be a carrier of the disease and a source of further infection. Therefore such cows are sold as barren for slaughter at much less than they are worth as milking cows. In this way the farmer's capital asset is dissipated and he no longer has the means to produce. Yet should he at this stage decide to put new capital into the purchase of disease-free cows, within twelve months of coming into his herd they will almost certainly have contracted the disease themselves and so the cycle continues.

Brucellosis in humans takes its name from an army doctor by the name of Bruce who first identified it in Malta towards the end of the last century and it became known as Malta fever. Now, however, it is a much greater health problem in Britain than in Malta. It is also, by virtue of its nature, known, but not quite accurately, as undulant fever.

I seem to remember reading something about this when I was a boy and, as far as I remember, the idea was that the problem was rife because they indulged in the practice of spreading human excreta on the land and that the bug was carried in the goats' milk. Whether or not there is anything in this I wouldn't know but it has always caused me to wonder what contribution is made to the people's health and well-being by the practice of an allegedly civilised nation in discharging increasing floods of crude sewage into its coastal waters. There must be some benefit because the idea seems to have gathered momentum since the people's eyes were opened for them in Conservation Year and direct outfall sewerage schemes have become the order of the day.

The symptoms of brucellosis are Protean. I know what this means because I looked it up after a doctor had told me so and I had said how very interesting. Proteus was an old man of the sea in Greek mythology who had received the gift of prophecy from the deity. But those who wished to consult him found him difficult of access for, on being questioned, he assumed different shapes and eluded their grasp. Likewise the symptoms of brucellosis in humans are, to coin a phrase, many and varied. There might be any one or a combination of violent sweating, headaches, debility, constipation, diarrhoea, vomiting, orchitis in men or mastitis in women, aching joints and limbs and a permutation of troubles from neurosis to irritability and suicidal depression. But that is only to touch on the problem. It seems that to date more than two hundred and thirty different symptoms and signs of brucellosis in humans have now been listed. It is, therefore, not easy to diagnose, for generations has been unsuspected, and can only be confirmed by blood samples. And even that can be confusing. It can affect the liver and it can be a killer. And where it doesn't actually kill, it can hardly, with its propensity for hanging around for generations, be conducive to longevity. Opinions amongst doctors vary as widely as opinions amongst vets. One doctor told me that the blood test for brucellosis in humans is conclusive. Others say this is not so.

Without going into too much technical detail, which in any case, I would not be competent to do, it seems that our brucellosis is slightly different from the Maltese variety. Theirs, which is endemic in the Mediterranean countries, is Brucella Melitensis and conveyed in goat's milk. I believe I am right in saying that cases in this country are rare. Ours is Brucella Abortus, which is the bovine strain. Although in humans the symptoms are

similar to undulant fever the brucellosis in humans in this
country is neither, strictly speaking, undulant fever nor Malta
fever. The brucellosis which we know is contracted from contact with infected animals and their discharges, from contact
with infected carcasses and, less frequently, from drinking
infected milk.

On the question of infected milk there is often a certain
amount of loose talk, mainly in the sporadic warnings of medical
officers of health or public health inspectors who, not infre-
quently, are no more helpful than they ought to be. The fact
is that possibly more than ninety-five per cent of the milk
retailed in this country is pasteurised and completely safe for
human consumption as well as health-giving. The record of the
milk producers' service to the community is a proud one and it
would not be an overstatement to say that on many farms the
hygienic conditions under which milk is produced are superior
to those which pertain in some hospitals. Anyone who has
seen a bandage roll its length along the floor and then be picked
up, re-rolled and placed in the medicine cupboard ready for
future use will know what I mean. Over the small amount of
milk sold by producer-retailers and which is not pasteurised
there is in theory a vigilant control and this milk, coming from
cows which are in brucellosis accredited herds, is normally
also perfectly safe. Even so, for reasons which can be discussed
later, the odd sample can always slip through the net. And, in
any case, other infection than brucellosis, such as salmonella,
can be carried in this way, so that it is only right to state the
firm belief that no milk should be sold for human consumption
in this country unless it has been pasteurised. Admittedly it
can be made safe by boiling, which will also destroy much of
its goodness, but who wants to drink boiled milk ?

For too long, the farming community have swept this nasty
business under the carpet in cases where their own herds have
been involved. Not surprisingly, the public at large has not

realised the extent of this hidden danger, particularly in the rural areas. But if mankind's history were a story of refusal to pass on the message from experience gained the hard way we would still be living in caves. The whole story of human progress is one of people benefitting from others' discoveries and learning. If the country were to learn more of the facts about brucellosis they might feel more inclined to make the money available to eradicate it. It is not a difficult task. It only needs money. I can tell them how to do it and it isn't the half-baked way the Ministry of Agriculture have gone about it so far.

Until the facts are brought fully and fearlessly into the open there can be no hope of any real progress. When I was a boy, tuberculosis, or consumption, were dirty words which were whispered gloomily and forbodingly from behind the back of the hand. Nice people didn't talk about such things. Eventually enlightenment came and the sanatoriums (or should it be sanatoria) were emptied.

The same applies to cancer, so often referred to as an incurable disease, a phrase beloved of local newspapers. When my wife underwent an operation for cancer twelve years ago in a Manchester hospital they said, "Now for goodness sake talk about it, so that people will know it can be cured." Very recently she was having an examination and a doctor new to her case history said, "What's all this about then?" "Only cancer" she said. "Oh, that's all right then" said the doctor "as long as it's nothing serious." It is probably fair to say that she has lost far more sleep worrying about the implications and ravages of brucellosis than ever she did about the cancer. Yet how often do we still hear of people who die from cancer just because they "left it too late". And they left it too late because they were scared by all the talk of an incurable disease. This is at last being overcome and great progress will result, and is indeed already resulting, because of it, in spite of the ravages of uninhibited diesel fumes and self-inflicted smoke inhalation.

And now it's high time to bring brucellosis out into the open. There is a greater danger to people than is realised because there is far more of it in cattle than is realised.

I shall make no attempt to write about things I don't understand. Not for me the talk of rising titres and agglutinins and blood cultures and of leucocyte counts. If the technically-minded want a book like that they must go and write one themselves. I can only write my own book. And all I know about is trouble. And practical trouble at that.

Chapter 2

BUILDING A HERD

IF the facts I give about my own experience are to have any real significance, and be fully understood and appreciated, then it is necessary at the outset to give something of the background. I hope such details will not be considered irrevelant, and indeed believe that my own feelings on many aspects of this miserable business will all be seen to have been influenced in one way or another by what I set down about my own farming and the writing which has developed from it.

Our herd of Guernseys was founded by my mother in 1938, originally as an adjunct to a private hotel, and, like Topsy, it "just growed". (My mother, in fact, died in childbirth when I was born and, in speaking of my mother in these pages, I am, of course, referring to my step-mother.) One of the two original cows was called Topsy and the other was Jeannie. They were two of the best cows that ever went into a cowshed and they were bought from a son of old G. H. Llewellin of Red Hill, who must have been one of the first people to bring Guernseys into Wales. We lived then at Amroth Castle in Pembrokeshire. My father was a builder and he and my mother, who had both been born on small-holdings, had taken it over, together with the small home-farm, on a fully repairing lease when it was as near derelict as makes no difference. My father, with his own workmen, was able to restore it at less cost than would have been possible for the estate who wouldn't even find a new gate for a tenant anyway.

My wife and I took over in 1945, just after my mother died, and, shortly afterwards, were able to take some more land from the same estate. Apart from running the catering side of the business our main concern was the building of a pedigree

Guernsey herd and we augmented the small herd that was there with a few careful purchases. It was a day of very high prices in the Guernsey breed, when characters with little knowledge of farming, but a real interest in tax allowances, were waving their cheque books about with considerable abandon to the ultimate detriment of the breed, as more and more good cattle got into the hands of those who didn't know what to do with them, whilst those who could have done something with them were left standing at the outside of the ring unable to bid. It dealt the breed a severe blow from which it is only now beginning to recover. I believed then that the Guernsey was a grand cow and now, more than a quarter of a century later, and in spite of all our bitter disappointments, I am more convinced of it than ever.

At that time the Guernsey herd book was closed but subsequently the English Guernsey Cattle Society introduced a grading-up register through which it was possible to bring the progeny of non-pedigree stock into the herd book as pedigrees. The purchases we had made ourselves were all pedigree but there were two unregistered cows, descendants alike of the two original house cows. One was June Rose, a grand-daughter of old Jeannie and the other was Topsy Dill, the only daughter which Topsy ever bred. She had six calves and Topsy Dill, her last, was her only heifer calf.

These two cows went straight into the grading-up scheme having qualified with figures way above the necessary requirements. The scheme provided that foundation cows, having qualified, should be inspected and would then enter class A. Provided the next three generations qualified on production the fourth generation would again be inspected and the progeny be eligible for registration in the herd book.

The first cow we submitted for inspection was June Rose. This was a great cow and, all these years afterwards, people ask me do I remember her. As if I could ever forget. The man who came to inspect her for the Society was that gentleman by birth and inclination, the late Gerard Cobb of the famous Toadsmoor herd. When he saw June Rose he just couldn't stop talking about her or asking me about her background. Then, as if that were not enough, he insisted on going all the way back down to the house, which was a longish walk, and said to his wife, "Phyllis, you really must come up to have a look at this cow."

Some time before that the late W. G. (Willie) Owen, of Braishfield in Hampshire, a famous cattle judge and Dairy Shorthorn breeder, had seen her as a heifer and said that he had never seen anything to compare with her. When Gerard Cobb passed her he said, "Well, if that's what you can submit for inspection for foundation stock I hope I'll be alive to see what you have by the time her descendants reach pedigree status." And we had hopes, because we were concentrating on bulls from the famous Chalvington herd and keeping to the blood of that great sire of the breed, Rose Lad 2nd of Maple Lodge.

I think, too, that we were perhaps on the right lines. June Rose's production figures spoke for themselves and showing her was only a formality hopelessly discouraging to other breeders. We only showed locally and at county shows and I just can't reckon how many championships she won. And I think we must have been on the right lines with her progeny. Early in the 1960's her grand-daughter, June Katie, won the championship at the Teifyside Show. Jim Jackson, vice-chairman of the Milk Marketing Board, was there. Having a drink in "the tent" afterwards—and there is an atmosphere about this part of the proceedings at this show so typical of the warm-hearted people of Cardiganshire—he said to me, "I like your cow, Roscoe. I'd like to buy her off you," When I laughed it off Jim Jackson said, "I'm serious, you know." Then he drew me to one side and said, "I'll give you £150 for her." At that time Guernseys had dropped considerably in price and it was a fair offer. But, looking to the future, I put temptation behind

me and said "No". We also spoke of the dangers of people
offering to buy your best stock and of the folly of refusing a
good offer because they will probably die anyway. This, of
course, is just a foolish farming superstition and there is no
more to be said about it except to mention that only weeks
afterwards June Katie slipped and spreadeagled herself on the
cowshed floor, damaging herself internally and was dead in a
few days. But by that time our much bigger trouble had already
started, although we didn't know it at the time. And I am going
ahead of my story.

Apart from this particular loss we still had a number of very
good animals of this family but were not so lucky with the
Topsy family who eventually died out. Even so, I have very
warm memories of the first Topsy, who was a "character", and
of her daughter, Topsy Dill. Topsy Dill, in fact, was a bit of a
bitch and I remember writing a piece about her on one occasion.
She was a real boss with beautiful upcurving horns which were
her crowning glory. I had always believed that a cow's horns
told you something of her quality and didn't care to go along
with the idea that was then gaining in popularity to have cows
de-horned, although, of course, we had to come to it in the end.
For a time I experimented. Some farmers, I knew, had put
little lengths of rubber piping over the horn-tips. Then I came
across brass knobs which could be screwed onto the tips of the
horn. The idea was that when the cow found she could no
longer do any damage she would no longer "bush" (a Pembrokeshire word meaning to gore) the other cows because she knew
that if they retaliated she would be helpless. With Topsy Dill
it worked like a charm. From being a complete tyrant over the
rest of the herd she was transmogrified and stood in the corner
of the field out of the way of the rest of the herd all day long.
So if you've ever wondered what transmogrified means you'll
know it means standing in the corner of a field with brass

knobs on. But in mentioning my writing I am going well ahead of my story again.

There were a few other foundation cows to which I would like to make some reference because I believe it will lead to a better understanding of what I shall try to say later on, about the problems with which we were eventually faced.

One of these was a cow called France's May, bred on Guernsey, the fountain head of the breed, and one of the last heifers to be imported from the island before the war and subsequent German occupation. When we bought France's May she had had six calves and had proved difficult to get in-calf again. We bought her privately in the hope that we might be lucky. If we could get her in-calf she would obviously be a marvellous animal from which to breed. She was the real type of fine-boned and deep-bodied cow typical of the best of the island animals. Had there not been a problem or element of doubt no money at that time could have bought such a cow. And, as it turned out, we were lucky. We got her in-calf, at the third time of asking, to our own Chalvington bull and, joy of joys, she had a heifer calf. The calf is worth a reference if only to save others from similar folly. It thrived at first and then became ill. The vet nearly lived with us for about three weeks and, right to the end, was completely mystified. When the calf died he opened her up and the penny dropped. With inadequate buildings we had reared the calf in a little shed adjoining, and under the same roof as the one in which we kept the T.V.O. for the tractor. The concentrated fumes from this fuel had been too much for her and her liver was absolutely rotten. Then the vet recalled health problems and losses in sheep in the Pembroke Dock area during the war when the oil tankers there had been bombed and the countryside blackened for miles around by soot from the fires which followed. What the future holds as a result of more recent oil developments at Milford Haven, and particu-

larly the fume-belching electricity generator, remains to be
seen.

The story, however, had no immediate sad ending. The cycle
having been re-started, France's May then proceeded to produce
another half-dozen calves. She produced thirteen altogether,
gave well over 100,000 lbs. of milk in her lifetime, milked on
for about three years in her last lactation and lived until her
twentieth year. Then, faced with the inevitable, we had her
shot and buried on the farm. We had always said she would
never go off the place alive and somehow couldn't face the
thought of her being fed to the hounds. Her descendants were
prominent in the herd until the end and at least one bull from
her has made a considerable contribution to the breed through
what is now one of the best herds in Wales, the owner of which
came to us for some foundation stock when he was starting.

Not all the purchases were as successful. When trying to
found a pedigree herd an older cow, which has perhaps had
some sort of accident, is often worth consideration for the sake
of what can be bred from her, but there is always an element of
risk. A couple of other such useful old cows which we bought
at the same time as we bought France's May produced nothing
at all. Then it came to 1950 which, as far as laying the founda-
tions of our herd was concerned, was "the big one". That year
saw the dispersal sale of the great Chalvington herd in Sussex.
By this time we were using our third Chalvington bull in a row.
I know it sounds ridiculous but my mother had gone there for
a bull in the first place because the cow in the advertisement of
the Chalvington herd in the Guernsey handbook "looked like
our Topsy". It was a most fortuitous likeness. To my way of
thinking there has never been a Guernsey herd to compare with it.
I was as much a fan of that herd as any schoolboy is of George
Best or Barry John. And the prices at the Chalvington dispersal
sale clearly demonstrated that I was not alone in my thinking.

The critics will claim that little was heard of the Chalvington animals after the sale. All I know is what marvellous results all three bulls produced with their stock in our own herd in strict and rigorous commercial conditions with the cows wintering outdoors, mainly because of shortage of buildings, but proving their hardiness in the doing of it. And, of course, the money-bags were much in evidence at the dispersal sale with few of the animals getting into the hands of people who were likely to be able to make any sort of mark with them, but some of them did, and Chalvington blood is still highly prized by leading Guernsey breeders.

By that time the prices of Guernseys had already started to fall considerably. Even so, all records for the breed were broken and stand to this day. It was not, by present-day standards, a big herd. There were twenty-six cows and served heifers, together with seven empty heifers and three bulls. The whole lot, pre-decimalisation, averaged £1,339/6/8 apiece.

We went to the sale, of course, and I had studied the catalogue and knew every detail in it better than Billy Graham knows his Bible. And we went more in faith than hope. We were interested in two of the bulls. The one was bought by Sir Henry Price of the Fifty Shilling Tailors and the other for Lord Beaverbrook. "Not" as they say "in the same league as us". But buying for Beaverbrook, of course, was Sandy Copland, managing director of Beaverbrook's Cricket Malherbie Estate, whom I did not then know but who has since become a very close friend of mine. The bull was Chalvington Bella's Roseboy and in later years I was to buy a bull calf from Cricket Malherbie out of one of his daughters.

There were but five heifer calves for sale. The first one was knocked down for 320 guineas, the bargain of the day. The next one made 400, the next one 600 and the last one 500. The first one, Chalvington Gaybelle, was a full sister to the bull we were then using, Chalvington Belle's Sunrise, and the record book says she was knocked down to one W. H. R. Howells.

This, it was fully intended, should be the last of our purchases. Circumstances, however, alter cases, We were at that time going over from hand-milking to a milking machine. Some of the cows simply rebelled and over the course of the next year or two we just had to sell some of them. And I'm not interested in the views of those who will say it was not necessary and we should have done this or that or the other thing. My grandfather always used to say "Everybody knows what to do with a kicking horse except the man who's got the bugger". At that time we had the best cowman we ever had. In fact he and the other chap with him at the time were the two best we've ever had. They were both with us ten or eleven years, both left only to go into more remunerative and less tied employment than farming, and both have remained good friends of mine over the passing years. They are the type which farming can ill-afford to lose and yet is losing every day because of the shameful unfairness with which the nation treats its most important industry.

We had no thought of buying-in cows to replace those which we sold, being content to wait for some very nice heifers of our own breeding to come into the herd, when we were hit by what seemed to us at the time a terrible calamity.

The first thing we had done in 1945 was to have the herd tuberculin-tested. There had been a few reactors, which we sold, and the herd soon became attested. The plan was already under way for the three counties of West Wales to become a clean area. From then on, of course, we had no trouble and had no reason to expect any. Then, out of the blue, we had our annual routine test in 1953 and had four reactors, all of which, of course, had to be slaughtered. And one of them was Chalvington Gaybelle. For weeks we puzzled and sought to find a possible source of infection until an auctioneer, who was in the know, asked me why I didn't enquire into the case history of my neighbours. They, it turned out, had had twenty-six reactors, and the trouble, it was said, had come from a cow with a tubercular udder that had been used for nursing calves.

At the time of what seemed then to be a major disaster I became pretty conversant with the implications of tuberculin testing in cattle and sources of infection. Now that there is talk of eradicating brucellosis there are people who compare the two problems and the two eradication schemes. To do so is to demonstrate beyond argument that they don't know what on earth they are talking about and their views can be put into a thimble and thrown away without being missed or making any contribution to the argument.

Sad though it all was, it was the sort of hazard to which farming is heir, and is accepted, and faced up to, and then you start off again and just go on doing the best you can. As a result we bought in three cows from dispersal sales, one of which was an old Chalvington cow, Primrose Garland. The Primroses were perhaps the best family in the Chalvington herd. Garland left one daughter, as did Chalvington Gaybelle before she had to be slaughtered, so that, in addition to some of those to which I have already referred, we still had every promise of good things to come. Those were the last females to be bought into the herd until the real trouble came and which will be the real subject of these pages.

We moved from Amroth and the herd moved with us, in the spring of 1960. How we came to leave there is a long story, far stranger than anything in fiction, which I propose to make the subject of a book later on, but which has no bearing on the subject under discussion.

Chapter 3

WIDENING INTERESTS

HAVING already made reference to my writing it would be as well to show here, as briefly as possible, how a combination of events and circumstances led to this becoming a full-time occupation.

It began, I suppose, in the autumn of 1945. I had only been back in Amroth a short time when I had a visitation from a deputation of some of the village boys who wanted me to teach them to play football. They didn't know whether they wanted to play rugby or soccer, and what was the difference, but they had been told that I had been a bit of a footballer, so that would be all right with them, as long as I could teach them.

Having established that they were but ten in number, I decided it would be easier to make eleven out of ten than fifteen out of ten, and opted for soccer. Although my own playing days had been almost entirely with the handling code, I had been born and brought up in Saundersfoot, just along the coast, where the village soccer team of the 1920's and early 30's had players who had become a legend in their own life-time. So I fancied I knew something about the game. Enough, at any rate, for these rudimentary purposes. When a couple of the older and more enterprising members turned up with newly-cut larch poles suitable for goal posts, I refrained from enquiring whether they had bought them or been given them in case the answer had been neither.

In no time at all I became chairman, manager, coach, chauffer and, of course, landlord, because I also had to provide the field, which all too often meant being groundsman as well. Before the end of it I had a little card to show that I had passed an examination to become a referee, I had managed to get a Junior

Division of the Pembrokeshire League started, and finally became chairman of the League's Management Committee. All because a handful of village boys wanted to play football. And, of course, I had been the official reporter to send in an account of the matches to the local paper.

It was perhaps inevitable that I should have become involved in this way, as I did in the life of the parish generally, because my roots went very deep into the area. My ancestors in direct line on my father's side had lived and moved and had their being in Amroth parish for two hundred years to my certain knowledge. My grandfather had, in fact, been a typical collier in the now defunct coal industry of the parish, with a useful little smallholding to go with it. There were scores of such holdings when I was a boy. It was there that I became familiar with the process of crushing great slabs of cattle cake in a hand-turned crusher, and where I first milked a cow by hand when I could scarecely have been much higher than the milking stool.

Had the involvement outside of my own affairs stopped at this it would not have been so bad but, in the course of the next ten years or so I also found myself serving on the District Council and finally became Clerk to the Parish Council. There were other time-consuming activities, too, not least of which was serving on the County Executive of the National Farmers' Union.

As I became more involved with the N.F.U., and meeting succeeded meeting, I became more and more incensed at the disgraceful treatment farming was having at the hands of the nation as politicians allowed, and indeed encouraged, a purblind electorate to believe that food subsidies were being paid to help the farmer rather than to keep down the cost of living to the consumer. As succeeding generations of dishonest politicians of all parties had instilled into the people the idea that cheap food was not only a possibility, but theirs as of right, there was probably nothing else they could do. Purely in an effort to promote the interests of the farming cause, I began, in 1954, to write a weekly column in our local paper. It was published then in Narberth under the title of the *Weekly News*. Later it was amalgamated with its sister paper, the *Tenby Observer*, and is now published in that town under the name of the *West Wales Weekly Observer*. But to those of my generation it will always be the "Narberth Paper".

The column appeared under the pen-name Barn Owl. The barn owl, of course, moves by night and is rarely seen. It was some time before anybody knew who the writer was. The barn owl is also a symbol of great wisdom and is likewise a good friend of the farmer. It was the first writing I had done since leaving school and, although turning out a weekly column was sometimes a chore, I enjoyed it and felt I might have been doing some good.

Nor was it without its more exciting and rumbustious moments. When, as I shall explain later, the column came to an end, the editor-proprietor of the paper, Glyn Walters, told a friend of mine that it had taken much of the zest out of his life as he found it odd not to be walking in fear of an action for libel coming round the next corner. But the column has not been forgotten and, all these years later, whenever Lord Netherthorpe (the former Jim Turner) sees me, he says, "And how is the old Barn Owl getting on?" Likewise Lord Woolley and Sir Richard Treharne still greet me as the old badger in remembrance of a later column which has also now ceased to exist. But that is something else which is in the future as far as this chronicling of events is concerned.

More immediately concerned with our own affairs I was conducting a one-man campaign with the local dairies to persuade them to pay the recognised premium for Channel Island milk, which they eventually did, and I was also making a great deal of effort to persuade the English Guernsey Cattle

Society to form a regional association for Wales. In addition
to the farm we were still running Amroth Castle and, although
we had given up catering and turned it into self-contained
holiday flats, there was still plenty to do.

In the spring of 1957, after a strenuous winter, I went down
with bronchitis and pneumonia and was very near to a nervous
breakdown. Probably the greatest contributory factor was the
fight I had been having for some years as a member of Narberth
R.D.C., who were promoting a sewerage scheme for Saundersfoot, and were determined, with the active support of the Clerk,
to make it a direct outfall into the sea. I loved Saundersfoot,
where I had been brought up, and I lived in Amroth so would
be affected by the appalling filth which would be washed
straight onto our beaches. I was even more appalled by the
apathy of various local bodies. My cousin and fellow representative for Amroth, Ivor Howell and I fought against it
every inch of the way. The St. Issells' Parish Council in neigh-
bouring Saundersfoot were as disinterested as if it had been a
proposal for somewhere in the north of Scotland.

As Clerk to the Amroth Parish Council I went to the public
enquiry as the main, and only voluble, objector to a scheme
being promoted by an authority of which I was also a member.
It was later established that, before ever the enquiry took place,
the inspector taking it had had a long session with the Clerk
and consulting engineer and that he had asked them to submit
plans for a direct outfall. The enquiry was no more than an
an expensive farce from the moment it opened until the Minister
finally put his rubber stamp on the Inspector's recommendations.

The people of Saundersfoot eventually saw the danger and
looked to Ivor and me whilst their own councillors waffled.
Eventually we were joined on the R.D.C. by a new councillor
from Saundersfoot, Tom Stickings, who was of the same
mind as we were.

The fight was not over. It was only beginning. As Clerk to the Parish Council I was able to pursue the matter with some diligence and the chief line of attack was through the good offices of the county M.P., Desmond Donnelly. He was then a Socialist and is now a Tory, having been something of his own denomination in the transitional period. I know nothing about politics and care less, but I do know that the ordinary rank and file voter never had a better individual member or one who worked more faithfully than Desmond worked in our case. Ivor and I went to see him and, when he had heard us out, he said, "All right, Roscoe. You go home and put it all down on paper and send it to me." I wrote for days on end and produced pages beyond number.

Desmond acted as the "clearing house" and sent my screeds to the Welsh Office in Cardiff. The characters concerned wrote back to Desmond and he sent their letters to me. Then the procedure started all over again, and so it went on. The case was at last re-opened just before my health broke down and the R.D.C. eventually had to install a treatment works.

The affair had played on my mind and I used to lie awake for hours at night turning it all over in my mind and wondering what step to take next and what points to make in future letters.

When I was ill in the spring of 1957 and came out of a brief spell of delirium and realised that the doctor had been to see me at an ungodly hour in the morning unbidden, and purely as a result of what he had seen the night before, I came to the conclusion that I must have been pretty ill. One whole day was unaccounted for. It was as good a time as any to take stock, and when the doctor said, "I take it you know the answer," I said yes, I knew the answer all right.

The position, of course, had become too ridiculous for words. I was chairman of the Pembrokeshire League, chairman of the Pembrokeshire N.F.U. Milk Committee, Clerk to the Parish Council, and member of the R.D.C. Up to that time I had not

missed a meeting for seven years of the N.F.U. County Executive and I had not missed a meeting of the R.D.C. since I
had been on it. The same applied to committee meetings and I
had never claimed a penny for expenses. I was chairman of the
Village Hall Committee, chairman of the Pembrokeshire
County Branch of National Milk Records somewhere around
that time, and was up to my eyes with the concert party for
whose sketches I wrote much of the script. Just as if that were
not enough, I was also a member of Tenby Round Table and
was writing my weekly Barn Owl column.

Having paid due regard to what the doctor had said, I felt
like Wee Georgie Wood on his twenty-first birthday when he
said to his mother, "Now then, mother, you've worked for me
for all these years, now you can go out and work for yourself".

Having resolved to concentrate on my own affairs and to
pack up everything except the N.F.U., my wife said I ought
to make some contribution to the life of the parish and so I
kept on as Clerk to the Parish Council. The rest just went by
the board, either immediately or in the course of the succeeding
months.

In doing something for myself I intended, amongst other
things, to write a book about Reuben Codd of Skomer, the
island off the west coast of Pembrokeshire. I had just made a
start when, in the July of that year, Sylvan Howell, who was
the P.R.O. for the N.F.U. in Wales, telephoned me fairly late
one night and said he wanted me to take on a full-time job as a
journalist.

Chapter 4

THE BLIND MILE

I SAID at the outset that this book is about brucellosis, and so it is. It will be seen later how this involvement as a journalist put me in a position to be able to discuss the problem with many different people over a wide area. It was an opportunity denied to many farmers and, because of my own close involvement, certain things meant more to me than to the ordinary run of agricultural journalists.

The job Sylvan Howell had in mind for me was with a new paper, due to come into being in September 1957. He had nothing at all to do with it but had been asked if he could make any recommendations. By that time the Barn Owl column had been running for three years and he was familiar with it. He also knew my loyalty to the N.F.U., and this was of great concern to him. I felt that, although I would have to abandon thoughts of my book for a while, I could find much of the time to do the job by utilising the time which I had previously been devoting to other things. In short, I would be doing something for myself. First of all, they wanted me to write a weekly column on the same lines as the Barn Owl. After some thought the pen-name I choose for this was Ben Brock and it had a little sketch of a badger in the top corner. Brock, of course, is a well-known countryman's name for the badger. This animal, because of the senseless persecution to which it has been subjected over the years, tends to be shy and move by night. He is very wise, very lovable, most amiable until roused and does far more good than harm. He is blamed for many misdeeds which are not of his doing by fools who misunderstand him. It is sad that the very ones who persecute him most are those whose friend he is.

The paper was based on North Wales and, in addition to writing my weekly column, I covered South Wales. I had to contact various reporters in the area who could cover different events on lineage, and wrote farm features and various reports myself. I also covered major events outside of Wales such as the national shows and certain important meetings in London. It all brought me into contact with a host of people connected with agriculture from many different areas and I made a wide circle of friends which included many agricultural journalists. I enjoyed the job.

We moved to our present farm in the spring of 1960. Although we left Amroth with many regrets we had been through a period of such harassment that we looked forward to life with a new zest and confidence on a farm which was our own. Although moving just over the border from Pembrokeshire into Carmarthenshire, we were only five or six miles from Amroth and could remain in touch with old friends. We still go to chapel there. I remained a member of my local branch of the N.F.U. and continued to serve, as I still do, on the Pembrokeshire N.F.U. County Executive.

The farm is a hundred acres on the red sandstone strip which runs eastwards from the south of Pembrokeshire. The house was just what we wanted and, although the buildings were old, they were substantial and adequate to our purposes, at any rate for the time being. There were two cowsheds, with ties for twenty-six cows and there was room to extend this to another dozen or so. As the herd increased we thought we might do this, to be able to tie forty cows, and instal pipe-line milking. All the propaganda being churned out by the National Agricultural Advisory Service during the last couple of decades has been for specialisation, usually in dairying, and for the dairy herd to be increased and housed in loose-housing and milked through a parlour. This farm, under such conditions, would be expected to carry anything up to a hundred cows with all replacements being bought in.

This was not our idea at all. We aimed to rear every heifer calf born into the herd and to buy nothing. We came here with twenty odd cows and had a number of heifers in-calf and about to come into the herd. It would not be long before we again had surplus stock to sell and there was never any problem about this. When we had anything to sell we rarely had to take it to the mart and were invariably able to sell it privately without going off the place. There had always been a steady stream of people wanting to know whether we had stock for sale.

The land is dry, easy to work and almost flat enough to be called a "one-horse farm". The road runs through the middle of it and we aimed to keep the cows on the side where the house and buildings are, and the heifers on the other side. We also planned to base a sheep enterprise on that side and to grow barley whenever it suited us from year to year. The grazing for the cows was laid out in a very compact and workable system of twenty-four paddocks, and the ewe flock was able to graze these in the winter, tidying up anything left by the cows. It was a most flexible system which suited us and the farm admirably.

The sheep we had that first year came from Reuben Codd, who had just finished his farming activities on Skomer Island, and had nowhere to put them. I had by that time written the book about him and it was published the following year. We kept the sheep for him that summer and bought them off him later. We then decided to go in for a flock of Llanwenog ewes. These black-faced sheep, traditional to north Cardiganshire and south and mid-Cardiganshire, appealed to me very much. There is no need to write much about them here except to be permitted to say that we won the cup for the best small flock on two occasions and that I was deeply touched when the members, some of them almost monoglot Welshmen whose business meetings were conducted in Welsh, elected me, an English-speaking Welshman, to their management committee and, subsequently, President of their Society.

The Guernsey herd, however, was to be the main prop to our farm economy with the emphasis on breeding. Our milk went to the same dairies, still at a premium, as it had at Amroth. There would be additional income from the sale of reliable stock and the occasional bull. Although Guernseys had dropped in price, and pedigree was beginning to be thought of as something for a small band of specialists, we saw no reason why, with our outstanding bloodlines, we should not be amongst such specialists in our own breed and, although the reward has always been for producing quantity, we had faith in the future of a quality product. Every animal we brought with us was home-bred.

There are some incidents in life and little sayings of other people which, somehow or other, seem to stick in our minds. One such was something which old Dan Howells, a great hill-sheep character who farmed at Ton Mawr in Glamorgan, said to me when I was writing a feature on him, He was talking about his own experience in having moved to another farm on the same range of mountains as the one on which he had always farmed. "But it doesn't matter" he said, "how similar the land and the conditions are. For the first year or two on a new farm it's like travelling a blind mile."

It is certainly true that conditions can vary from farm to farm, let alone from one area to another. It is the sort of thing which puts the home-bred animal at a premium. On some farms there seems to be a greater danger of cattle being poisoned through eating fern shoots or ragwort. On some farms it is red-water. In our own case the first hazard round the corner whilst travelling the blind mile was when we found a beautiful young heifer stretched out, stiff and cold, in the field, having died, so it was established, from blackleg. In more than twenty years with animals it was the first case I had seen.

So, injections of calves for blackleg before turning them out in the spring became part of the routine. The loss of one heifer was not too high a price to pay for a little learning provided there would be no repetition and, indeed, we have never lost anything since from this cause.

When the summer came the cows were not milking very well. Even fresh-calved cows were disappointing. Then the odd cow began to cough and the vet, at last, said "I'm afraid you've got husk."

I said, "What ! In the cows ?"

"Yes" he said, "in the cows."

This was a new one to me. I had had a little experience of this pneumonia type of infection but we had never been troubled with it at Amroth. It is caused by a worm picked up at grazing and which affects the lungs. Normally it is associated with young stock.

When I asked him how serious it was in cows, he said, "For your own sake I could wish it was something more simple and straightforward like foot-and-mouth."

I said, "What the hell are you talking about ?"

"Well" said the vet, "if you had foot-and-mouth you'd have to slaughter the herd and you'd get compensation and a lot of sympathy. But with this lot you'll get neither sympathy nor compensation."

I saw his point when, by August, we had most of the cows in the cowshed by day and night and were feeding them on hay. Fortunately we had plenty. We have sold a bit of hay from time to time but have not had to buy any since we have been here. But that first year we lost two thousand pounds. A few of the cows' lungs were so badly damaged that they never got over it properly. That winter, dosing of the calves for husk was added to the routine and we have not had a case since.

So much for the blind mile. There is an attitude amongst farming people at such times as these which prompts them to say "Never mind. As long as it's outside the door we'll manage." That winter, however, it didn't stay "outside the door" for, right in the middle of it, my wife went into hospital with that so-called incurable disease known as cancer.

Like the old lady who was never heard to complain about the cup that was half-empty, but always gave thanks for the cup that was half-full, we, too, have much for which to be thankful. We have joked about it since, but lived under a dark shadow that made life seem black at the time.

As far as cows are concerned they say that where you have livestock you must expect to have deadstock. And, although there were no immediate casualties, there were problems. This is what farming is all about. All we could do was keep faith and keep plodding on, for once you lose your faith you have lost everything.

My wife's cancer had been taken in time and the herd had got over the husk. The blind mile had been travelled and things began to improve. The cows were settling down, we had good heifers coming on all the time, and earlier promise looked like being fulfilled.

Chapter 5

THE HIDDEN DANGER

How brucellosis first infected our herd, or whence it came, I still have no idea. I am even willing to admit that I cannot say when it came, partly because it had obviously been present long before it made its biggest impact and partly, perhaps chiefly, because we knew virtually nothing about the disease. I am writing now with the benefit of hindsight. Had we known then what we know now, we would have got out of cows straightaway. For years we had been living in a fool's paradise and took a little time to get wise to the facts of life. As I tried to stress in the first chapter, farmers have never been very willing to talk about abortion troubles in their herd. They will tell you if they lose a cow calving, or from milk fever, or if she does something stupid like getting her head stuck in a piece of machinery and breaking her neck. But abortion is a different story.

It is perhaps understandable, therefore, that when trouble came we had to find out the hard way. Unfortunately, it was also the slow way, and that made it harder still.

For years contagious abortion had been a dread disease in the dairy herd. People who were farming in the 1920's and 30's have some bitter memories of it. I heard a great deal about it as a boy but saw nothing of it at first hand. I knew of one farmer who went into his cowshed one winter morning and there were twenty-six dead calves in the gutter from cows that had aborted in the night. That was a big herd in those days. So it was a big loss. Except that, as the old country saying goes, "It's only those who've got 'em can lose 'em". But if you lose only six, and happen only to have six anyway, then it would be difficult to try to persuade you that your loss has not been so great. Nothing can be any more than total.

The first attempt to control the disease in this country with vaccine was with a vaccine known as M 45/20. This was unsatisfactory, however, because it was a live vaccine which could cause infection and find its way into the milk. The Ministry of Agriculture, therefore, withdrew it and in 1944 introduced, from America, another live vaccine known as S.19.

There can be no question that, for some years, this vaccine did a very useful job. For one thing, it enabled cows to carry a calf full-time, so that, perhaps for the first time for a long time, a farmer could look forward to a live calf and a milking cow. To people who had suffered, this was Nirvanah, and S.19 appeared to true believers as next best thing to the second-coming.

It was not long before the farming community was lulled into a false sense of security, and a new generation was born which knew nothing of the dangers and of the trials and tribulations of those in whose footsteps they followed. They heard about it, of course, but it didn't really mean anything. It is only when you have had trouble yourself that you are really willing to listen. There are farmers today, good friends of mine, who are convinced that I am some sort of crank who takes things too seriously. But there are not as many now as there were when I first started writing about it and they thought I was really off my nut.

It so happened that when I did hit trouble I was willing to talk about it. All my life I have believed in standing up to be counted. But, because of our unwillingness as an industry to talk about this problem openly, we have remained very much in the dark on something which is vital to our very exsitence. The only ones willing to talk about contagious abortion are those with clean herds and no trouble, often dear old gentlemen who will tell you what it was like in the days before S.19.

Venerate their white hairs and their wisdom and respect them
for the struggles through which they emerged triumphant,
albeit in some cases with a little timely boost from Adolf Hitler,
but don't ask them about contagious abortion in today's conditions. When I started writing about brucellosis in 1965 all I
knew was that it had hit us, but I still had a great deal to learn.

Having said this much it is perhaps as well to look more
closely and critically at what was happening. To start with, a
vaccine does not cure a disease. It only prevents it. In many
cases it does not prevent it, but holds the worst effects at bay.
By enabling infected cows to carry a calf full-time, S.19 was
merely blanketing down the disease whilst helping to keep it in
being. It was with us without perhaps being very much in
evidence and we had soon reached the stage when, if a man was
having trouble he was a fool and it was his own fault for not
using S.19. In the very early days it was understood that the
animal needed to be blood-tested after vaccination to see whether
the vaccine had taken or not, but this idea soon went by the
the board because of the cost and the trouble.

The Ministry have been the chief manufacturers of S.19 in
this country and, although there was a small charge for its use
at first, it eventually became a free service for anyone prepared
to use it. The ideal was considered to be to inject calves at
between three and six months of age, but adult animals were
also vaccinated, especially where there was trouble, and it was
not unusual for vets to recommend a boost by further vaccination after a cow had had her third calf. The trouble is that, when
animals are vaccinated after the age of puberty, there is the
complication that they will show positive on a blood test, with
it being impossible to tell whether the cow is infected or not.

The vets have learned about these things as they have gone
and we, the farmers, have been the guinea pigs. I have no
objection to this. It is aimed at helping us and, if it so happens
that it helps the nation at the same time, so much the better.

But there is still so little known that you can talk to a dozen different vets and get a dozen different opinions. Yet it is not their fault. The fault lies in the firmly entrenched system whereby, even the Ministry vets, with practical experience in the field, are not in charge of the business. They are consulted, of course, and they advise, but the scheme is administered by faceless wonders in the seats of power in the civil service who know as much about problems in the practical field as a pig knows about a holiday. It is happening in every aspect of our national life, and it is a system which, if it is allowed to continue, will eventually destroy society and even civilisation. Turn where you like and you see the crippling effects of the dead hand of the civil service. Nowhere is it more evident than in the case of brucellosis in this country, and in our failure to come to grips with the problem of its eradication. The people are the losers all along the line as these characters jostle for promotion and appear with monotonous regularity in the Honours' Lists.

Whatever else has been discovered or not discovered, however, about the whole complex business, it has been found, and some farmers have found it out the hard way, that S.19 is by no means one hundred per cent. Its advocates point out that no vaccine is, and they claim that S.19 is seventy per cent effective. The truth of the matter is that, if the real facts could be established, it would be found to be more like only fifty per cent. And even that guess could be on the generous side.

What, then, are the sources of infection? The real concentration of infection comes when an infected cow aborts or calves full-time, and the fluids, the placenta, the aborted foetus or the full-time calf, whether dead or alive, are all heavily contaminated. The real danger period is, therefore, for maybe a few days before the event and until the cow stops discharging maybe a few weeks later.

The chances are, particularly where there has been no previous case on the farm, that the first abortion comes as a surprise and happens in the field. It can happen very quickly, literally over-night, without any visible warning to enable the farmer to take action. This means that the infection is everywhere about the pasture and the whole herd is in danger, and farmers, at any rate, will be aware of cows' remarkable propensity for licking the calf of another cow and sniffing around all that goes with it. Point out that virtually all infection is through the mouth and the horrifying picture begins to be seen.

It is, however, much more sinister than this and there is involved at this stage another of the hidden dangers with which the business is fraught and which, until recently, has not been declared, even if it has been understood. The concentration of infection in an infected placenta is many times greater than the measure of infection which S.19 will withstand, so that the amount of protection when trouble comes is slight indeed. Nor must the farmer with a clean herd think that he can sleep easy of nights just because he has double fencing all round the boundaries with his neighbours, which happened to be one of the requirements in the initial stages of eradication of tuber-culosis from the national herd. It only needs a dog or a fox to drag an infected placenta over the hedge onto a clean farm and that's that and all about it. When we were in the depth of our trouble we had a heifer abort one night out in the field. In the evening, after milking, she had been perfectly all right. By the morning she had aborted with a seven month calf and was fussing and licking round the spot where it had happened. But there was not a sign of the calf to be seen. We went over the fourteen acre field with a toothcomb, but all we found, after hours of searching, was part of a leg two hundred yards away. There was something sinister and frightening about it.

There is no need for me to spell out how foxes have increased and what a menace they have now become, but it might not be a bad plan to draw attention to the part played by birds such as the magpie. If there is any carrion or filth about, they are much in evidence. More lambs are lost and ewes killed by having their eyes picked out by these pests than ever is generally realised by those who have never kept sheep. Likewise if there is an aborted foetus, or a placenta, or any discharge in a field, they are on the spot. And I have begun to wonder whether more troubles than have been suspected have perhaps come from this source. Without carrying any infected material from one farm to another they can eat it in one place and spread infection in their own droppings elsewhere.

Then, of course, there is the question as to how infection at birth will subsequently affect the calf, and whether infection picked up in suckling a cow will remain with it for the rest of its life, what effect it will have and whether it can be passed on from this source. There is much that is not yet known. Nor, would I hazard a guess, has enough been done to ascertain whether some new mutant of the known types has now appeared. One vet tells me that there are nine known brucella types in this country and S.19 is partially effective against four of them. Another vet tells me this is not so and that there are only three and S.19 is effective against all three. He also says that I'm way off beam talking about this unidentified mutant business. Yet there are vets who are seriously concerned about this possibility.

There is the further belief that S.19 became unreliable early in the 1960's, when it was kept by a method of freeze-drying, but that it is now once again as effective as it is ever likely to be. So the arguments and the discussions continue.

In all this sorry business, however, there is one point on which there can be no argument at all, and that is that probably the biggest single contributory factor in the upsurge of brucellosis has been the swing towards the method of so-called live-stock husbandry under which farmers have been increasingly obliged to manage their milking herds. And nobody was more guilty of advocating and hastening this trend than the Ministry's own National Agricultural Advisory Service. At demonstrations, meetings, conferences and discussions up and down the country they have preached the gospel of the need for the big to get bigger. When I started on my own we reckoned, if I remember rightly, to milk eight cows an hour by hand. If you could milk fourteen in that time you were really in a hurry to get finished, and somebody was likely to cast some doubts on whether you had milked them all out clean or not. Then the milking machine took over and it was a great boon in a herd of thirty cows, where one man could do the whole job a good bit more easily than three or four people had done it previously.

It was not long before we became familiar with the gospel that there was something sacred about being able to sack another man, whilst bemoaning the fact that everywhere the unemployment question was becoming more serious, and the only increase in numbers was in the civil service. The men left on the farms, whilst earning little more than they could get on the dole, and far less than they could get in other and easier occupations, were told that they had to milk increasing numbers of cows on their own. I think about a hundred cows per man is the present going rate. Or is it a hundred and twenty? We shall certainly not see many such cowmen mature to ripe middle-age, let alone old-age, in their master's service on this stunt the same as we did in days of yore.

All along the line cowsheds have been knocked down, abandoned, or converted, whilst milking parlours, and all the trimmings concomitant with such a system, have mushroomed. Cow kennels, cubicles, covered yards, self-feed silage and cows wallowing in each other's discharge have become the order of the day. Out in the fields intensification has been seen as the only means of salvation. On farms where twenty cows were once reckoned to be a useful herd they are now considered to be understocked if they only have sixty. Paddocks and strip-grazing have become standard tools of management, and each cow has been forced to graze nearer and ever nearer to the excretion and discharges of her herd mates.

At the same time there was an almost unamimous clamour from the same Ministry advisers to persuade farmers not to do anything so old-fashioned and inefficient as to rear calves for their herd replacements, but to buy-in cows to maximise the output from the conserved fodder by producing as much milk as possible. No word was ever mentioned about what disease could thus be spread from farm to farm.

Milk production is an exacting business, both physically and in financial investment, and the fact that farmers were prepared to go along with this trend only emphasised the complete lack of balance in the rest of our farming, because few men in their right minds would opt for such a life if they could possibly hope to make as much out of beef, or sheep, or corn, or anything else you care to mention. The fact that it was a producers' organisation in the shape of the Milk Marketing Board, which had to struggle to market the huge increase in supplies of milk as advantageously as possible, was no concern of the N.A.A.S. with their five day week, six weeks annual leave (some of the more hard-done by have to settle for as little as three), annual increments, up-grading and superannuation. It is also an unhappy lesson, which has been learned the hard way since 1947, that if the country has been short of any particular food commodity, the surest and quickest way to get an increase has been to reduce the price of it to the farmer. What fools we must be.

Small wonder, in the face of such a trend, that stockmen have been unable to give their charges the individual treatment so essential to good management. When things are going all right there are no insurmountable problems, but once trouble comes, all hell is let loose.

I have already said that a cow can be infected and still carry her calf full-time. Sometimes in such cases the cow does not cleanse properly after calving and this would be a job for the vet three days later. I have also mentioned the question of infertility and the loss which this involves. To start with there is the vet's bill for examining the cow and washing her out. The cost of this can be added to the loss, when the cow is finally given up as a bad job and sold barren, at a fraction of what she was worth on last year's balance sheet, when the farmer has failed to get her in-calf. This is where contagious abortion is the real killer for the dairy farmer, more than the loss of the calf and the milk, because the unit of production itself has been lost along with the capital which has been dissipated.

It, therefore, behoves a farmer who finds he is having cows with retained cleansings, or an undue number having to be sold as barreners, to start looking into things very closely because this could be the first sign that he has contagious abortion in his herd.

Unfortunately, these were things I did not know when they began to happen in our own herd. There was far less open discussion about brucellosis in the early 1960's than there has been of late.

Chapter 6

PROMISE OF BETTER DAYS

THE recovery from the effects of the husk was not immediate. The odd cow would give trouble and one or two had been so badly infected that they were never really any good again. For the most part, however, a cure had been effected and things were improving. Indeed, before we were even out of this particular wood, a representative from one of the commercial firms came to ask us if he could have some of our calves for a calf-rearing demonstration which they were staging in Carmarthen. Whether it was because of their breeding, his feed, or our management, I wouldn't know, but they were certainly thriving and looking well.

As heifers which had not been affected by the husk came into the herd there was still more cause for optimism. We had, too, been doing quite a bit of work with paddocks and the grass was doing well. The previous owner had taken out a considerable length of hedging on the house side of the road and, although this was something I would not have done, it at least had the advantage of giving us an unencumbered area in which to set up a system. With the help of John Davies, the N.A.A.S. grassland officer for Wales, we were able to establish a most convenient and successful lay-out of twenty-four paddocks which proved to be a great help in the management of the herd and of the grass. It was doing so well that one of the N.A.A.S. grassland men, who was making a film on paddock grazing at the time, took a considerable number of shots of the grass and different aspects of the set-up. Narberth Grassland Society, too, of which I am a member, decided to have a look, and those who came seemed to be quite impressed. We were also paddock-grazing our ewes and lambs on the other side of the road with a forward creep system and this was working well.

It was about that time that the editor of *Welsh Farm News* asked John Davies if he could give him a few names of farmers who would be worth writing up in a grassland feature which he was planning. The N.A.A.S., of course, have been a constant source of such information for members of the agricultural press. John Davies rather surprised him, I suspect, by asking him hadn't he heard of Roscoe Howells at Cwmbrwyn and told him that we had just about the best grass he had seen that summer. I was not too keen on the idea but, having written so many features on others' farming, I could not really object to somebody writing about my own. It was finally decided to ask Reg Evans, the highly respected director of the Ministry's Experimental Farm at Trawscoed, where he did such marvellous work, and Reg agreed and duly obtained the necessary permission to write this feature for publication.

It was, I am willing to admit, a great morale booster to know that such a shrewd and able man, who had such a complete grasp of so many aspects of Welsh agriculture, should like what he saw. He particularly liked the way we had been able to integrate our dairy and sheep enterprises to help towards a balanced farm economy. But the only part of his feature which I would choose to quote is where he concluded by saying, "Of course Mr. Howells will maintain that he is the perfect example of an absentee farmer and that all credit is due to Mrs. Howells, his able lieutenant, and to his cowman and part-time shepherd".

To be able to quote this I had to search for a long time recently to find the old paper which carried this feature and, when I found it, I could almost have wished I hadn't bothered, for it also carried a picture of "a promising bunch" of sixteen in-calf heifers. To see that picture and its caption now is enough to make a man weep. A few months later Sandy Copland stayed the night with us. He had been speaking at a forum arranged

by Channel Island breeders in West Wales, and somehow or
other the date had been very conveniently fixed to coincide
with the visit of his beloved All Blacks to Cardiff, so that he
could go and see this team (which made few friends by the
miserable rugby they played on that tour) on the way back to
Somerset. I can always remember this very down-to-earth
colonial looking at these heifers and saying, "My goodness,
Roscoe, if ever you wanted to raise some quick cash, you've
got some money's worth there". Three years later I was able
to tell him that the whole lot had aborted and/or gone off barren
since he had seen them.

It was at this time that the Guernsey Breeders' Association in
Wales held its first herd competition for members and we were
close runners up to the winner. Some years previously we had
done a not very wise thing in going outside of our chosen
blood-line to bring in a bull with a very high butterfat ancestry.
He certainly made his mark with a spectacular increase in the
butterfat of his daughters but they lost out considerably on
type. With this bunch of heifers by another bull coming on,
however, we could see that we were again on the right lines
and that it would take quite a herd to be placed above us in
another year or two.

Without knowing that we had already run into trouble, and
could have been the cause of spreading disease, we were still
selling a heifer occasionally whenever someone wanted to buy
privately, and we also supported a small sale of Channel Island
cattle held in Carmarthen. It was a terrible day with ice and
snow, which had a serious effect on both attendance and trade
but, for what it was worth, we had the satisfaction of making
top price.

As something by way of a little side-issue I might mention
here that, having seen a great deal, as a journalist, of other
people's farming and the sort of milking-parlour set-up beloved
of N.A.A.S. demonstrations on efficiency, I had almost con-

vinced myself that, with our two separate cowsheds and bucket milking plant, we must be amongst the most inefficient farmers in the country. To establish the truth, the whole truth and nothing but the truth, therefore, I decided to join up with the new-fangled Low Cost Production service being operated by the Milk Marketing Board.

Surprisingly, gratifyingly, but not very convincingly, this bit of nonsense showed us to be very efficient indeed. One month, in fact, we were the most efficient producers in the whole of our very efficient code-numbered but nameless group. The whole thing, of course, was straight from cloud cuckoo land, because they put in hay at some sort of cost of production, and barley the same, with no sort of compensatory factor between this and the figures of the man who was buying a great deal of his feed. I didn't want to know whether we were making hay efficiently, but whether we were inefficient milk producers and, on this point, the service left us in blissful ignorance. I decided that here was the way to save a few pounds to become more efficient by opting out when the M.M.B. had a ridiculous display board at the Dairy Show, based on some of these hope-lessly unrealistic figures, and telling the world how ignorant farmers could help themselves if only they would take advantage of this forward-looking service.

If the idea has been given that life by now had promised to become a bed of roses I should dispel that illusion immediately. As is inevitable on any farm, there were the usual problems and disappointments. For some time we had considerable staffing problems, which could not be made anything other than more difficult when my journalistic work took me away a fair part of the time. And, even though the job had the great merit that I worked from my home, where nearly all my writing was done, there is a limit to how much on-the-spot farming problems can be regarded as contributing to inspiration or be conducive to the writing of flowing prose. Indeed, this coincided with the paper itself going through a very bad patch with some of the problems having their own particular repercussion on me.

The bitter weather during the winter of 1962-63 was another serious problem and it is doubtful whether anybody could really put a figure on what it must have cost the industry. Memory plays funny tricks, and opinions vary, but I have always believed that it was a worse winter than that of 1946-47. Our cows wintered out, to the great surprise of neighbours who had imagined the breed to be delicate, and came through it well, but it was a horrible time, with frozen pipes, difficulties of milk collection, and a hundred and one other things which can make life more burdensome under such conditions.

To make matters worse, we had planned to lamb a month earlier in the spring, and we had a hundred and twenty ewes on the place.

There were more problems and disappointments to come when, our labour problems having been overcome, the very good cowman referred to in Reg Evans' article had a breakdown in health, as a result of which he had to give up farming for a time. Not long afterwards, the part-time shepherd referred to in the same article, a grand boy and real good friend, lost his father and had to stay home full-time with his brother. By the time we again had another good cowman, a first-class stockman whom I had known since he had played in our boys' football team, we had run into our brucellosis troubles, so that he never had a chance to show what either he or the herd could achieve. He was with us for five years until we were in sight of giving up the unequal struggle but, with the best will in the world, it can never make economic sense to be paying a man up to thirty pounds a week in a small herd, riddled with abortion, and from which the very important revenue through sale of surplus stock has been removed. The fact that he is capable of earning it is neither here nor there if you cannot provide him with the

wherewithal to do so. It was a great disappointment to both of
us, and there were times when I think he felt it as badly as I did.

For reasons which I shall explain later, I am not writing this
from records. I keep a very inadequate diary in which I mention
odd things on different days. It does not give anything like a
complete picture of the farm, but serves as a reminder of the
odd event. To quote the farm records with the dates and
numbers of every infection and abortion, and the names of the
animals involved, would be wearisome and pointless. Such an
exercise would only be of any use to anybody carrying out a
fact-finding survey.

I have said earlier that I could not really say when brucellosis
hit our herd, and that I could not even say exactly when I realised
and admitted to myself that we had it. Having shown something
of my attitude to getting on top of the blackleg and husk
problems, by injections and dosing, it will come as no surprise
to learn that we had used S.19 vaccine as a matter of routine
right from the start. There is the occasional abortion on most
farms and, disappointing though it is, it is not necessarily the
signal for gloom and despondency. All abortions are not contagious or caused by brucellosis. They can be caused by any
one of many things.

Our first abortion was in the autumn of 1963 and it was, as
is so often the case, one of the cows which figured prominently
in plans for the future. It was, in fact, a cow from the Chalvington Primrose family, which we were very keen to establish, and
it was a heifer calf, six weeks premature and dead. Contagious
abortion was not even considered, especially when we had been
subjected to a particularly heavy spell of low-flying by airborne
jet-propelled lunatics who are virtually impossible to identify.
We had also been somewhat troubled with night-time shooting
by poachers, which again is not calculated to induce chewing
of the cud by contented cows, to say the least. In December we
had a heifer with a premature calf. She was one of the family

of the Toadsmoor cow, which was one of the three we had bought in 1953. In this case the vet said the calf was mummified so that seemed to account for that.

When we had two more somewhat premature calves we had blood samples taken and these were positive. From questions we asked at the time, however, we began to suspect, and have since had ample evidence to believe, that these tests are about as reliable as a politician's promise. The trouble then seemed to clear up, and small wonder, because we had always used S.19. This was why we had used it. One vet said that I need not lose any sleep about it as long as we had been using S.19. The trouble, he said, would soon "work itself out".

We had no further abortion until the autumn of 1964 and that again, when it came, was no cause for serious alarm because the cow in question had broken into a field of grass which had been dressed with basic slag, and this gave her severe diarrhoea, and she was straining and under the weather generally. Yes, of course, this sort of thing could, and very often did, cause abortion. Unfortunately there were a few more which now came with slightly premature and dead calves. There was a need for action, and what better than the prophylactic balm of that modern miracle, the eighth wonder of the world, S.19 no less. A policy of boosting in the cows was decided upon, but not, mark you, without misgivings from the Ministry veterinary people, who gave their permission grudgingly. On the other hand, their objections are on the grounds that it gums up their precious blood tests. It is also claimed that it has been found that it is not a very effective procedure anyway. I can second that motion with great feeling and speaking from first-hand knowledge acquired too late.

Chapter 7

A DYING PIG

HAVING accepted that the impossible had happened and that we really did have contagious abortion, we began to look back with a more critical eye at the milk registers and cowshed diaries of day-to-day events. A picture soon unfolded of an uncommon number of veterinary visits to remove retained cleansings, and to wash out cows that were returning to service. Let me repeat, it is easy to be wise after the event. At that time we simply did not connect this sort of thing with brucellosis. It took even a number of abortions for us to grasp that it had happened and to realise, with that horrible sinking feeling, that S.19 was not what it was cracked up to be.

A rugby full-back can have a lonely and empty old feeling in the seconds before he goes down on the ball in front of a pack of rampaging forwards. Likewise a tail-end batsman can die many deaths as he faces really fast bowling with only a handful of runs needed. For that matter, some pretty horrible pictures and possibilities can be conjured up by the attitude of un-sympathetic bank managers in the hours of need. But such feelings are as nothing to the realisation that contagious abortion is the enemy. The other things will pass, but with brucellosis who can tell ?

By the end of the winter, however, things seemed to have returned to normal again, and we were comforted by the assurance that it would "work itself out," and by the additional protection of the boost with S.19. But by now the retained cleansings and returning to service were beginning to tell their own story. More and more cows were going off barren. Even a cow which had calved normally could still have been infected and the side effects would be infertility. We must have had the disease on the place for a couple of years. By the summer of 1965 we were down to twenty cows from the thirty five which the herd had numbered before all the trouble began.

In the autumn of 1965 I did what I should have done much earlier and asked for the Ministry to be called in. Before I write another line I would like to say that, throughout our trouble, the Ministry's County Veterinary Officer could not possibly have been more helpful and sympathetic. Equally, I am bound to say, at other levels I detected more concern at protecting the tarnished image of S.19, and defending the Ministry's paucity of thinking on the subject, than ever there was for me and my problems.

It was at this stage, as Gerard Hoffnung has it to be in his classic, the Bricklayer's Lament, that I must have lost my presence of mind. But forgive me. As a journalist, I had covered Lord-only-knows how many demonstrations arranged by the N.A.A.S. and some of the commercial firms with their technical advisers fresh from college. Words beyond number I had faithfully quoted on the need to maximise the old output. I had listened for hours to erudite gentlemen hold forth upon the subject. It couldn't possibly be that they were all wrong and that I was the only one in step. And there is a Pembrokeshire expression to the effect that a dying pig will kick. It seems a pity to go down without a fight and we certainly couldn't expect to do anything else but go under if we only had twenty cows and they were further depleted into the bargain. Remember, too, that I had by this time become established as an agricultural journalist and it was more important than ever that I should not now forfeit this income. I mention this because there is the thought that the man on the family farm, by tightening his belt, might have been able to plod on. But, because of my journalistic work, I had to go on employing a man and a boy. And, employing them, I had to find the money to pay

them. So, as I said, forgive me. I decided that we would have
to buy more cows.

Again, in all fairness to the Ministry vet, he advised against
it, saying that as we now appeared to have weathered the storm
it could be as well to wait for calves and heifers to develop and
come into the herd. If we bought in, he said, there would be a
danger of the whole thing flaring up again. But I quoted the
facts of life as propounded at the N.A.A.S. demonstrations and
conferences. Furthermore, anything we bought in would have
to have been vaccinated with S.19. Like Bernard Miles and
his square on the Pythagoras, it was unanswerable.

The idea was to buy in up to about twenty-five cows to make
it a viable proposition. If all went well we would then have to
think in terms of a milking parlour, more convenient handling,
and an increased number of cows to produce more milk, to earn
more money, to pay for the sort of set-up necessary to keep
more cows, to produce more milk. Obviously any thought of
breeding had to be cast firmly into the back of the mind for a
long time. There was no longer any thought of selecting only
the best heifers to come into the herd. We were just thankful
for anything that calved normally. From here on much of the
interest went out of the job, and breeding and families and
individuals were forgotten as matters of survival itself occupied
our minds.

I think now that it would have been better at that time if we
had bought in cows, not put the bull on them, milked them out
and sold them barren. With nothing in-calf on the place the
brucellosis would automatically have died out. But there were
drawbacks to this plan. It would have depended for its success on
our being able to buy fresh-calved Guernsey cows at a price
which would not have shown too great a loss when they were
sold twelve months or so later to the butcher. The Guernsey
cow would not be the most suitable for this sort of policy, nor
would it be all that easy to find as many as we wanted for the

purpose in this part of the world. In that case, it may be argued, why confine ourselves to Guernseys? For one thing, if we had not done so we would have lost our valuable and hard-won Channel Island premium. And the plan, in any case, would have involved selling our own cows for slaughter. When I say our own cows, I mean our life's work, everything we had hoped and worked to achieve in farming. Obviously we had the thought in our minds, and always fostered the hope, that one day we would see these troubles behind us and be able to think again in terms of breeding. And I've told you before what gramfer used to say about the kicking horse.

Don't let us upset ourselves by talking about money and the raising of it. Let us rather stick to the facts of the case. We bought in fifteen point-of-calving Guernsey cows and heifers, nearly all of them privately, and there was only one dud amongst them. And that was my own fault, for I bought her by 'phone without seeing her. Otherwise they were all honest animals and good value for money. They milked well and, according to L.C.P., whose services had not then been dispensed with, we were right up in the picture cards when it came to the old efficiency. They were all blood-tested and had all been vaccinated with S.19.

Top priority then became to ensure that as many cows as possible were in-calf for the following autumn, because the Channel Island premium puts the emphasis on winter milk. In the flush and eternal promise of spring-time we reckoned up the score and kept our eyes open. By June time the picture was becoming more clear and we were able to say that twenty-four cows, the new fifteen and nine of our own original animals, had been bulled to calve from Sempteber to December 1966.

Again, I shall not go into wearisome details. The first two aborted whilst I was away covering the Royal Show at Stoneleigh. The third one aborted the day after I came back from covering the Royal Welsh. In the event, out of the twenty-four, six calved down normally, eight aborted and ten were sold off barren. I may not have mentioned earlier that, when this wretched business is in being, cows that are not in-calf will often show no signs of coming into season, and the vet cannot tell you until some months have passed. Yet when I began to write about these things there were stupid old goats who said that nobody should take any notice of people like me who had no practical experience of what they were writing about.

Troubles, however, are made to be overcome and, again as they say in Pembrokeshire, you must never let 'em think your mother bred a jibber. So then we embarked on what was intended to be a series of tests. The previous boosting with S.19 complicated things, of course, but interesting information came to light, even so, and some of it was really helpful. For example, many of the milk ring tests proved negative, some were interpreted as indefinite and some as suspicious. But one was Negative with capital letters and it said of the cow from which it was taken, "As she is known to have aborted twice she should be regarded as being infected". Perhaps I have faith in the wrong things in life but from then on I somehow lost interest in the testing side of it.

This particular cow, which had to be regarded as infected, was in fact the heifer of the Toadsmoor breeding that had, we now recognised, aborted as a heifer when she had the mummified calf. She carried her second calf but, whether because of, or in spite of, the boost with S.19, she had slipped her third. She also slipped her fourth and always gave enough milk to be worth her place in the cowshed. And by now, of course, there was no question of selling off cows just because they aborted. As long as they milked somehow they had to be kept on in the hope of something better next time.

In due course this cow carried her fifth calf full-time. By this time there was much talk about plans which the Ministry had for eradicating brucellosis and we were thinking in terms of weeding out some of the known and obvious aborters and, therefore, carriers. I put it to our vet who had not previously been on the scene, that, according to Ministry rules and regulations, now that this cow had calved normally, it would be perfectly in order for us to offer her for sale by public auction as a fresh-calved milking cow, once she had stopped discharging. The fact that she would probably calve down normally in the next herd to which she went and spread untold disease and misery seemed not to matter. He could do nothing other than agree with me that the whole set-up was criminal. In the fullness of time she was milked out and, like all our others, went off to the butcher when she went dry.

I am, however, going a little ahead of the story again because, by this time, I had begun to write about the subject with some feeling.

Chapter 8

"WE TESTIFY THAT WE HAVE SEEN . . . "

I OPENED what seemed to us to be my one man campaign with what I considered to be a very restrained piece of writing in my Ben Brock column of November 13th, 1965. By that time I felt I had had enough experience to know what I was talking about and pose a few thoughts for readers' consideration. The touch to the powder keg had come in the form of the hotting up of the Ministry's campaign to get everybody onto using S.19 in their much-vaunted free vaccination service.

I said, amongst other things,

"However commendable might be the Ministry's efforts to encourage the use of S.19 all round the place, it would be more to their credit if they also quite honestly warned the ignorant nit-wits who do not avail themselves of the offer that, even if they did, S.19 is a long way from being a reliable protection against abortion."

It was evidently far too mild and moderate a statement for readers who had become accustomed to much more full-blooded fare. There was little or no response. Obviously this uncertainty and feeling the way were no good at all. I needed to be much more sure of my facts and spell things out in words of one syllable. I felt I had reached this stage when I returned to the attack in August 1966. By that time, as I had forecast, it had become the fashionable thing for the multifarious advisers, all part of the burden which the industry has to carry groaning on its back, to deride the backward farmers who were failing to inject their calves with S.19 vaccine.

I therefore wrote,

"The opportunity to criticise a section of the farming

community is far more important to them than to know
what they are talking about.

Anyone who knows what he is talking about can tell
you that S.19 is perfectly harmless until such time as the
animal is exposed to infection. Once that happens it is no
longer harmless but is, nevertheless, completely and utterly
useless. Therefore, quod erat demonstrandum, S.19 is
useless.

The real smart cookies are the ones who have had nothing
at all to do with it. The real mugs are those of us who, for
the last twenty odd years, have used the stuff and fooled
ourselves that we were really progressive and leading all the
others."

Having referred to the fact that vets in private practice could
say little, because they were employed by the Ministry, and
discussed the implications of facing up to our problems, and
advocated compulsory slaughter, I said,

"The objection has always been that it was too costly.
Does nobody take into account what the disease is costing
the country? If the figure to keep the disease in being
could be worked out it surely wouldn't make the cost of
eradication look too dear.

I could for once in a way have been misinformed, but
I've heard it said that a pilot scheme was put forward
recently but it was turned down because the money was
needed to appoint another two hundred and sixty odd
blokes to the N.A.A.S.

Good God above. Don't they think we've had enough
advice and conferences and discussions and demonstrations?
Isn't it about time we looked for some practical help
instead?"

Having then referred to the announcement of a pilot scheme
for eradication, details of which were not then available, I
concluded by saying,

"Good luck to the scheme and may the best man win. But when it comes to cleaning up our herds the only real answer is for the Ministry to admit that S.19 is a failure, chuck all existing stocks into the ash bucket, and get to grips with a scheme that will really do the job.

If we can slaughter for tuberculosis and foot-and-mouth we can slaughter for this lot. As things stand, only the countryman's health is in danger through the possibility of drinking untreated milk. If the townsman's health were endangered it would be different. That would be bloody serious comrades."

I did not on that occasion spell out the danger to vets, farmers and cowmen from contact with infected animals, and I was in error, of course, in saying that only country people were in danger through the possibility of drinking infected milk. Large numbers of caravaners and visitors to the country generally run the same gauntlet. But more of that later.

Possibly what had niggled me as much as anything was a completely lunatic suggestion by Mr. Joseph Godber that all calves should be compulsorily vaccinated with S.19. He was then Shadow Minister of Agriculture and suggested a fine for anybody not complying. By this method, he suggested, in a couple of years everything in the garden would be lovely. Not that it matters what these types say. The same man made many promises on other occasions about what the Tories would do for agriculture when they returned to power, but, as is customary in the manoeuvring synonymous with this brand of deception, euphemistically referred to as electioneering, when the chance came, good care was taken to see that he wasn't there to answer for his utterances. Now that he has more recently belatedly come to the throne it will be interesting to see whether he has learned anything in the interim. But, if Mr. Godber was not there to answer for his utterances, Ben Brock was most certainly there to be shot at, and there was no shortage of takers.

The only reply at first was from an aged gentleman, long since retired, who mixed his diatribe on ancient history with an incomprehensible reference to donkeys, and it was obvious, he said, that I had had no practical experience behind it.

That week, however, several people met me, and others 'phoned me, and said now look here they knew I liked to start an argument but I had really gone too far this time and it was going to bounce back on me and do me terrible harm.

The following week, therefore, I started my column by saying,

"Not to put too fine a point on it, I'd say that it's reasonable to assume that a large number of normally very sound people are in violent disagreement with me over my lack of confidence in S.19 vaccine which I expressed in this column a couple of weeks ago.

So, just in case any of you may be worrying on my account, let me tell you straight away that I don't care a couple of monkey's uncles what they think. My only regret is that some of them are good friends of mine and it's on the cards that some of them could one day be due for a rude awakening."

I further added in the course of that week's dissertation that unlike the old gentleman who had written and was

"dreaming of bygone days, I am very much living with the problem at present and, whilst I would neither claim to be young nor a visionary (I had referred to the prophet Joel), I yet live in hopes of at least seeing some realistic action being taken to put an end to this plague. And if I can do anything in a quiet way to stir up some action on this score then, by God, I'll do it."

We then proceeded to have some foolish suggestions that it was only with the Channel Island breeds that S.19 didn't work, but there has been plenty of publicity about breakdowns in herds of all breeds more recently for it to be unnecessary to enter into further discussion of that particular point in these pages.

Oddly enough, and I had long become accustomed to having to defend myself when attacked in print, it was the letter, which subsequently evoked references to Delaney's Donkey and Paddy McGinty's Goat, and which had attacked me the previous week, which caused correspondents from all over the place to write to me personally and to support me wholeheartedly in letters to the editor. Some of the facts they quoted were not merely disturbing, but frightening. I shall not attempt to quote them here in great detail, nor much of what I wrote in my column during those weeks, because it would only be repeating some of what I have already said here. But I would like to quote in full a letter which came from a Mrs. E. Nabholztanner, of Binningen in Switzerland. She wrote,

> "My husband is a veterinary surgeon and I work as his assistant and secretary and so I hope I am qualified to write on Brucellosis and what I have learned about it during 21 years. Before the battle against Brucellosis began, about a quarter of all abortions were caused by it in my country. So, one could say, why so much fuss about it ? But no other cause gives such enormous losses and no other is contagious to humans. Not only untreated milk, but, even more, untreated cream and butter (most bacteria are in the cream !) are dangerous for consumers. Farmers, farm workers, butchers and vets can get the infection through small wounds in the skin.
>
> When the authorities in my country decided to get rid of brucellosis, we first had a go with live virus vaccine which soon proved to be too dangerous, and I remember very well the introduction of S.19. It seemed to be the solution of the problem and all were very fond of it. The plan was to vaccinate all calves in the country to build up a national herd of immunised cows. Therefore in the early fifties

every farmer throughout the country was obliged to have vaccinated every calf destined to be reared, and I remember well the many explanations we had to give to our clients.

The officials and the vets were very confident, but some years later we learned that the immunity given by S.19 was too weak to withstand a heavy infection.

The picture was the following : In infected herds we had less abortions, but we could not protect free herds from infection. Why ? In infected herds the cows build up an immunity within one to three years. Most of the cows slip the calf only once, others twice, seldom more times. So after one to three years you don't have abortions till the followers without immunity are pregnant.

This gives the undulating effect in infected herds—some years abortions, some years normal. If the calves in those herds were vaccinated with S.19 they often had enough immunity to withstand the infection (caused by secreting older cows) later. But free herds contacted with Brucellosis suffered heavy losses despite vaccination with S.19. Therefore the vaccination scheme was abandoned some eight or nine years ago and S.19 was forbidden.

Instead of this the eradication scheme took place and our country now is free of Brucellosis. The eradication scheme worked out as follows :—

1. Every herd in the country was blood sampled.

2. Herds proving free were obliged to separate bought-in animals until blood and milk samples had shown the animal free. Down-calvers had to be separated until afterbirth had proved free. From very early birth the afterbirth had to be sent for examination. Blood sampling was repeated every year.

3. In infected herds all cows were milk sampled and every afterbirth, even of normal calving cows, had to be sent for examination. Every animal secreting brucella

had to be slaughtered. Animals reacting only could be kept during the first stage of the scheme.

4. In the second stage of the scheme not only secreting animals, but blood reactors too, had to be slaughtered. Herds with more than 50 per cent reactors went to the butcher as a whole. In the other herds the reactors had to go. The remaining animals of the original herd had to be separated from the bought-in Brucella-free replacements until they had proved free on two examinations, to prevent reinfection by a remaining cow.

All expenditures for sampling were paid by the Government. Animals due to go to the butcher were valued by a veterinary officer and the owner received 80 per cent of this from the Government. As soon as an animal was found secreting or reacting it had to be slaughtered within nine days. Since our country has been free of Brucellosis every herd is blood sampled every second year. Twice a year a bulk sample of every herd is tested for Brucellosis. If a positive reaction is found every animal of the suspect herd will be sampled to find out the reacting one which is slaughtered. The afterbirth of every early birth has to be sent for examination.

On pastures (most of our rearing pastures are common ones), for expositions, for markets and for private trading no animal is accepted without a certificate that the animal and the herd where it is coming from is free of tuberculosis and Brucellosis.

I should like to support you in your fight for an eradication scheme. The definitive solution of the Brucellosis problem is not only important for the farmers but for the nation's health too. Try to get the help of the doctors and your member of Parliament. I wish you all the best."

Having quoted that letter which spells out the answer to the problem in language which, it would be supposed, even a civil servant could understand, I would think there is little need for me to say very much more about vaccination and what should now be done. At the time the letter was published I felt just a bit like a lamb with two mothers, because here was someone who really had long experience of a successful campaign saying much the same as I had been saying myself. The proletariat, however, appeared to remain singularly unimpressed.

Very shortly afterwards a report appeared in the paper about a discussion on the subject of brucellosis on a Welsh programme of the B.B.C. and said,

> "It caused some listeners surprise to learn that criticism of the S.19 form of treatment for brucellosis is only on a limited scale. If we heard the vet aright, he said it came from one farmer only and no one in the studio remonstrated at this observation."

I thought at the time that the programme would not be likely either to do much harm to the cause or make any worthwhile contribution towards it. Indeed the signs were encouraging Whereas, a short time previously, I had been merely an irresponsible journalist, I was now, it seemed, a farmer. I would be less than human if I failed to admit that when, much more recently, I spoke at a Rotary luncheon, it was a sweet feeling when the same vet was called upon to propose the vote of thanks and said, "A few years ago, on a certain other subject, a lot of us thought that Roscoe was talking nonsense, but now we have had to acknowledge that he was right after all."

There was another crumb of comfort at the Royal Dairy Show in October 1966, when I now had the bit well between my teeth. Reporting on the show under my own name I said,

> "At the press conference on the eve of the show further emphasis was given to the need for a more positive approach in the fight to eradicate brucellosis from the nation's dairy herds. Viscount Cranbourne, President of the Royal

Association of British Dairy Farmers, expressed the opinion that nothing would be a surer way of getting support from dairy farmers than to pay more for milk from brucella free herds.

He was quick to add, however, that this was purely his personal opinion and maybe this was just as well, otherwise he might have found himself with some awkward questions to answer as to what had been done by way of offering practical help and advice to those who were in trouble and were anxious to do something about it.

The association's own scheme, which was launched at last year's Diry Show, already has more people waiting to participate than can be accommodated because of the lack of laboratory facilities.

What questions were asked on the subject, in fact, were immediately passed to Mr. John Passfield, a veterinary surgeon member of the Council of the R.A.B.D.F., who has done much solid work in trying to focus attention on the critical position concerning this problem.

He said that it was cause for great disappointment that the Ministry of Agriculture had not yet announced their intentions in the matter and, as far as vaccination was concerned, whilst a dead vaccine might have something to commend it over the controversial S.19 live vaccine, the only real answer was for an eradication scheme which used no vaccine at all."

And all the Delaneys Donkeys and Paddy McGinty's Goats in the world couldn't say that this man didn't know what he was talking about.

At this time, of course, we were, as will be appreciated from what I have said earlier, right in the thick of our troubles, and I felt very deeply about what was happening. In the New Year of 1967 I was very pleased to learn that my writings on the subject had earned me a special commendation in the Fisons Journalist of the Year Awards, administered by the Guild of Agricultural Journalists, of which body I was the first Welsh member.

This award, whilst being very gratifying, didn't seem to make my, perhaps rather strongly held, views any more acceptable in other quarters. I particularly deplored the failure of the Carmarthenshire county branch of the N.F.U. to come to grips with the problem about this time. The monthly journal of the Union in the county carried an article on brucellosis by the Ministry's County Veterinary Officer. I was a member of the Carmarthenshire branch and sent a letter in reply which was published in the next issue. The Ministry vet was asked to comment on my letter, but when I again sent in a reply this was summarily stamped upon on the grounds that, as the Ministry vet had been asked to contribute his article in the first place, it would be wrong to allow any further comment on it. He would, I know, have been the first to welcome such comment, but, if we cannot look to our own Union to promote the views and interests of members, then where in the name of democracy can we look ? It was this sort of myopic attitude with its paucity of thinking which made it that much easier a couple of decades ago to set up a splinter movement under the name of the Farmers' Union of Wales. Fortunately, as I have said, I have retained my active membership of the Pembrokeshire county branch of the Union and through these rather more enlightened channels managed to make one or two views known, with, I hope and like to think, some beneficial results.

It seems incredible that Ministry progress reports on how the eradication scheme is working should be quoted in so many sections of the press as though they were statements of fact. So far the scheme has achieved nothing but a considerable amount of damage to the interests of our dairy farmers, yet week after week, we can read in our farming papers that the scheme is making great progress.   If such fatuous hand-outs were sent to newspapers on almost any other subject they would be thrown straight into the wastepaper basket but, because there is nobody on the staffs of these publications who really knows about brucellosis, they get by unchallenged time and time again.

Chapter 9

". . . AND YE RECEIVE NOT OUR WITNESS"

(THIS chapter is already somewhat outdated because it was
written early in 1972. In May of 1972 the Minister announced
certain new conditions and revised terms of compensation which
served to paper over some of the cracks in the creaking structure
of the eradication scheme. But because I have endeavoured to
deal with the subject in some sort of chronological order I think
it could serve a useful purpose to leave it as written because it
does at least help to show what has led up to the most recent
situation which can be discussed later.)

The Minister of Agriculture's long awaited scheme for the
eradication of brucellosis was launched in November 1967.
Not since the days of Horace himself had a mightier mountain
gone into labour and brought forth a more miserable mouse.
Had he been injected with the stalest possible sample of freeze-
dried S.19 vaccine he could not have been seen to have produced a worse abortion.

For some time there had been considerable pressure to bring
back the old M.45/20 which had now been "killed" and was
available as a dead vaccine. This was now done so that it could
be used in adult stock as an added measure of protection where
there was a breakdown in a herd.

Again there is no need to go into a vast amount of detail and
I shall content myself by showing what my own reaction to the
scheme was by quoting part of my Ben Brock column of
September 23rd, 1967. I said,

"In bowing to increasing demand and at last allowing
the use of a dead vaccine in the shape of strain 45/20 the
Ministry said in their recent statement that this is less

effective than S.19. There are very sound practising vets,
however, who dispute this statement.

Not that it matters, mark you, because the Ministry
also admit at last that S.19 does not give complete protection in the face of heavy infection. This has been said
a dozen times in this column and comes as news to nobody
who has concerned himself with the problem previously.

There are those who believe that the Ministry's announcement shows that they are even now only concerned about
the effect that S.19 is likely to have on their apology of a
scheme for eradication rather than for the lack of protection
in the herds which have been using it and have come into
contact with infection.

The encouraging thing is that the announcement acknow-
ledges at long last, if only by implication, that vaccination
can help to control a disease but never to eradicate it, and
there is at last the statement that, 'It is intended to bring all
vaccination to an end when that step becomes possible'.

The only thing holding this up, it seems, is lack of sufficient known brucella-free replacements which makes compulsory eradication by slaughter impracticable at present.

This, of course, is utter hogwash and clearly demonstrates
the complete cluelessness of those whose responsibility it is
to give a lead in this matter.

Nobody is talking of starting to slaughter today or
tomorrow. What is needed is a declaration immediately
that compulsory slaughter will start in four years' time and
that as from today a premium will be paid on every dairy
heifer calf reared. There should then be no shortage of
replacements when the time comes that they are needed.

Of course, this would involve a certain amount of organisation in segregation and management because the biggest
weapon in the fight against brucellosis is still scrupulous
hygiene and commonsense. But do the Ministry drive

this home in their recent announcement? Not one dicky-bird.

And this is where they have failed so shamefully in their duty all along, in shouting the odds about a highly suspect vaccine with never a mention of the basic essentials of management and stock segregation. They will say but, of course, they have always said this. But they haven't.

And if they had really given proper consideration to this aspect of the question is it likely that they would now be introducing a scheme which will not only allow, but encourage, people to sell infected cows in the open market from which they can go into the herds of the foolhardy, who think they are safe because they are using S.19, and where they will be able to wreak untold havoc? A hell of a good scheme, I must say. So good in fact it should forthwith be stuffed and presented to an ungrateful and unsuspecting nation."

That last senctence doesn't read quite right somehow, but at this distance of time I can hardly be expected to remember the exact wording which the editor would apparently have deemed prudent to modify slightly at the time, although we would undoubtedly have had a brief telephonic and vituperative disagreement on the issue.

I quote this reference to the scheme because it was made at the time of its announcement and it was how I felt then. To say the same thing now would only look like being wise after the event. As more and more details are given of herds in which there have been catastrophic breakdowns, it can be seen that much of the trouble has been caused by the peddling of this filth at the Ministry's behest in an overall set-up in which the economic pressures are all towards the keeping of more and more cows. Admittedly this gap has now been plugged, after incalculable damage has been done, because it has now been made illegal to sell a cow which has reacted for other than slaughter. But it is still possible for a cow to abort, not be blood-tested and then be sold when she calves again. And until there are more incentives in a sensible scheme there are plenty of farmers who can see the sense of not testing at all.

Even now, there is ample evidence that the lesson has not gone home, because early in 1972 a Ministry publication *Y Maes*, serving the three counties of West Wales, carried two articles on brucellosis which would have been laughable had they not been so pathetic and dangerous. The one writer felt that the only part of his offering which needed underlining was the importance of using S.19. The other, far worse, worked out sums showing the implications of the scheme based on the desirability of buying-in replacement stock. Talk about the left hand not knowing what the right hand is doing. I should think that if any A.D.A.S. man (that's what N.A.A.S. men are called now, but a rose by any other name) really wanted to know how many beans make five, almost any Ministry vet could tell him that the one thing they don't want to encourage is buying in.

In all fairness, it is only right to say that the Ministry men are not alone, for the clamour now being made by certain N.F.U. people is enough to cause despair. Of course, the scheme is a dead loss and, of course, it will never work. But not for the reasons they are claiming. On all hands there is talk about reactors and bloodtests. As far as I am concerned this is an academic irrelevancy. The ones to be winkled out are the aborters and carriers. Let us get first things first. The shouting is all about the hardship of people having to pay the high price of replacements in eradication areas without a thought being given to the suffering of those who have been crucified financially in the areas where the filth has been peddled.

Needless to say, nothing has been done to have a long-term plan for the breeding and rearing of dairy heifer replacements, and so a shortage of stock was inevitable. Hence prices have rocketed, and we hear the complaints because of the price of replacements. Of course, this is a hardship. But what of the man outside the area who is losing money through lack of production every day and, even if he could find the money, dare not buy in because of the certainty of a flare up the same as we had ? And this is what people seem to overlook when comparing the problem of eradicating brucellosis with the earlier campaign to eradicate tuberculosis. In the case of tuberculosis a man could continue in production without seeing his cows abort and going off barren. There was no urgency to buy-in replacements and therefore he could afford to plod on until the scheme included his particular area. With brucellosis the same man just has to die a slow death whilst waiting to come within the scope of the half-baked scheme we now have. The areas of the first eradication operation were completely inadequate. What was needed was for an area to include such as the whole of the three counties of West Wales and go to town on the carriers and aborters, and making it worth while not to replace them until things had levelled out. Instead, there have been insignificant pockets in which farmers have been encouraged to bring new stock onto their farms, whilst being surrounded by areas in which the disease is now getting a real foot-hold. They have been slaughtereing the less harmful cows in these small areas whilst storing up more troubles by allowing the real infection spreaders to run amok all round them.

In herds where there is trouble, outside of the eradication areas, a policy of using 45/20 in adult stock is now often pursued. Sometimes it works and sometimes it does not. If it does work it still only helps to keep the disease in being. It is expected of people in this predicament to do the honourable thing and send their cows for slaughter. These are the people who have already been hit hardest and are least in a position to do the right thing at such cost to themselves. These are the people to whom any eradication scheme should be directing its help.

They will tell you that we cannot start on the carriers in the dirty areas because we would not have the herd replacements, oblivious to the fact that a large number of services in the dairy herds are by beef bulls. If we suddenly switched to dairy bulls to ensure heifer replacements in four years time, they say we would be short of beef. Just as if there is no precedent to importing what we want or, for that matter and worse still, even what we don't want, anytime it suits the Government of the day to do so. We are already slaughtereing anyway, and getting precious little in return for it. No wonder a cloth-capped stranger who travelled in the same compartment as I did not long ago between Swansea and Port Talbot, whilst I was bound for London, waxed eloquent on some idiosyncracy or other of our national character, and declared, "We must be God's chosen people boyo, because nobody else could be so bloody stupid and get away with it."

From time to time the press faithfully carry reports that the eradication scheme is going well. They do so in good faith because this is what it says in the Ministry's hand-outs, and those who know what is happening are left to wonder whether it is a case of barefaced dishonesty or blind stupidity. All they have done is clean up a small number of clean herds in some very small clean areas whilst, in general terms, the plan, which was never more than a segregation scheme, is in disarray and a complete shambles.

Farmers themselves are sufficiently conversant with the financial details of the scheme for there to be no need to spell them out here, and town readers, I fancy, would not be unduly interested. The main point is that something far more realistic needs to be done and, whilst there are those who say we cannot afford to eradicate brucellosis, it is true to say that we cannot afford not to do so. This is not even to take into account the implications of a state of affairs in which we, the Stock Farm of the World, cannot sell to so many countries which refuse to

admit stock which has been vaccinated with S.19 nor to consider
how this will affect us once we are in the Common Market.

It is difficult not to feel a measure of sympathy for the Ministry
of Agriculture in this situation, because it is well-known, and
argued in their defence, that the real flies in the ointment are
the people at the Treasury. This, of course is only too true,
but who can say that the Ministry are likely to be able to put a case
forward when, at the same time, they are claiming that progress
is being made and that their scheme is a good one? The same
sort of defence has always been advanced when successive inept
Ministers of Agriculture have sold farmers down the river at
Price Review after Price Review. Yet what happened, was
it in 1955, when the Government had a majority of four, and
the Rt. Hon. Fred Peart, who was Minister of Agriculture at
the time, still allowed the Government, whose errand boy he
was, to impose its decision? Here was a case where he could
have got what he wanted for the industry if he had threatened
to resign but, in doing so, he would have wrecked his political
future. Not perhaps the ideal pilot for a scheme with the pitfalls
and responsibilities inherent in the need of ridding the nation's
dairy herd of brucellosis.

This then is the background behind all that led up to the next
insignificant bite at the problem which came with the announcement in May of 1972 by the Minister of Agriculture that there
was to be a revision of the terms for compensation for compulsory slaughter. And here, let it be said, I have considerable
sympathy with the Ministry of Agriculture and rather less
patience with my own organisation, the N.F.U.

All along the line the Ministry have dragged their feet, and
whatever they have done has been a case of too little too late.
Never do they seem to have done anything of their own volition
or in any way given a lead. Always they have seemed capable

of acting only after heavy criticism and waiting for other sections
of the industry to spell out the problems.

In this case, the new approach to the scheme, there had been
incessant demands from the N.F.U. for something to be done.
Unfortunately, the people shouting the odds seemed to be
concerned only with the question of the inadequate compensation paid to those who had to slaughter compulsorily and the
high prices they had to pay for replacements. Thus the Ministry
could be forgiven for thinking they had made a useful contribution and the N.F.U. dutifully welcomed the announcement.
But whilst easing the burden in some cases of hardship, the
scheme comes no nearer to eradicating the disease than it did
before.

When the new conditions were first announced I was greatly
encouraged to find reference to compulsory slaughter of dangerous contacts. I really thought they were coming to grips
with the problem at last and intended to winkle out the aborters
and known excrators. All it meant, however, was animals
which passed a blood test in a herd where a number reacted.
Such a herd would be slaughtered in its entirety in an eradication
area. But, outside the areas, nothing is yet being done to locate
the real culprits by making brucellosis a notifiable disease and
starting in the obvious place with aborters and known excretors.

This is imperative if any real progress is going to be made,
and compensation must be paid to make it possible for a man
to carry on without buying in replacements. Indeed buying
in replacements is the last thing on earth a man in any sort of
trouble should be encouraged to do.

Having said that the new scheme has eased the burden in
some cases of hardship, it is also true to say that many who tried
to do the right thing in years gone by are now finding themselves heavily penalised whilst those who bought any old cow
any old where are having a bonanza.

All over the country it is quite usual for accredited and nonaccredited animals to travel in the same lorry. Whatever the rules and regulations may say, it is a fact that cattle from a brucellosis accredited herd can go straight into a lorry after it has just been transporting a cow that has aborted.

Small wonder, with the danger of infection from foxes, dogs and birds being what it is, that those who administer the new scheme quietly admit to their disappointments. It is going very well they say. So many clean herds and so many in the pipe-line. A picture which, officially, is very encouraging. But ask them about the breakdowns. Ah, yes, the breakdowns. Most disappointing. And quite unaccountable.

There are none so blind as those who will not see. Nor will they be in the least impressed by anything in these pages. But the time is coming when the breakdowns will be even more numerous and it will have to be acknowledged that we are as far away from eradication as ever.

A farmer I know, not in an eradication area, recently made an admission to me. He is one of those who knows his way about. He had had some cows abort and, in an attempt to tackle the problem, had a milk ring test on the whole herd in batches of six. The picture was alarming. His own vet called in the man from the Ministry, and the farmer asked them what he was to do. The Ministry man, not surprisingly and not unreasonably perhaps, said, "The first thing you'll have to do now is blood test every cow in the herd."

And what do you think the farmer said to that? He said, "Not on your bloody life, boss."

Obviously, anything which reacted could not be sold for anything other than slaughter. As it is, how many of those same cows will, quite legally, find their way onto the open market to be sold as "a genuine cow, gentlemen. Right and straight in all quarters and a good milker. How much shall I say?"

The Ministry plugged the hole, admittedly, which they themselves created and by which they encouraged farmers to peddle the disease round the country. But don't let any one be naïve enough to believe that the door is still not wide open, whilst their edict that only stock vaccinated with S.19 will be eligible for compensation is about as helpful as putting red pepper on a gumboil.

At the same time as the most recent regulations were introduced it was also announced that the Government had accepted the recommendations of the Industrial Injuries Act Advisory Council to place brucellosis in humans on the list of prescribed diseases. Welcome though this is insofar as it will help people employed as cowmen, it will be no use at all to the self-employed, the family farmers, who will contract this wretched disease in the years to come.

Chapter 10

WHAT PRICE ABORTION ?

WHEN we caught our real packet, as I have already explained, in the summer and autumn of 1966, it was obvious that we would have to think again very seriously on whatever provisional plan we may have had in mind. It was no good thinking in terms of expanding on the herd as it was. We could have slaughtered the herd, allowed the land to remain free of stock for a few months, and then got going again with clean cattle. How easy it all is in theory. Apart from the fact that it takes no account of the complete loss of whatever capital assets may have been left in the potential of a breeding herd with some of the best blood in the country, and the destruction of a life interest, it fails to spell out where and how the necessary money is to be raised and repaid. It should also be remembered that to restart on such a basis would have needed anywhere from sixty to eighty cows, with the costly new buildings which would have been needed to house and handle such a herd. The lay-out of our buildings is such that a programme of cheap adaptation was out of the question and it would have needed a completely new set-up. It is doubtful whether such a venture could have been considered justified by any one so largely dependant on hired help, with so much money at stake and the constant need to be right on top of the job.

Having considered the business with all these thoughts in mind, there was the further realisation that, even if it could be made to look like a possibility on paper, there was the risk, indeed the virtual certainty, of reinfection from whatever the source may have been from which we had it in the first place. Even then, before the Ministry's left-handed so-called eradication scheme had encouraged the peddling of infection round the country's markets, I knew there was more of the disease about than was ever being openly acknowledged. Whenever I got up at a meeting and held forth upon the subject I knew that immediately afterwards there would be people coming up to me saying yes, of course, they had had some trouble, too, but they didn't like to say anything about it in the meeting. And when yet another article appeared people would telephone or write and tell me, usually in confidence, of their own troubles and experience.

It was obvious that the best we could hope to do was hang on. We had long since ensured that even cows calving normally should calve indoors. Previously we had allowed cows, particularly in the spring and summer, to calve outdoors. Many a time I've seen a cow pick her place to calve in a nice corner by the wood only to lose interest when brought in and then, on being turned out, go immediately to the chosen spot and produce a calf. But all this had been stopped and no chances could be taken. No cows could be bought-in, and there was going to be a long row to hoe if we were to hope to keep going. Cows which aborted and were giving some milk had to be kept on, along with those which had calved and were milking normally, in the hope that they might calve all right next time. There were, too, some promising heifers again coming along. It was one of those odd things but, whereas we had, I suppose, about a fifty per cent breakdown in the herd, those cows which did calve to full-time nearly all had heifer calves. Most farmers from time to time experience a long run of either bull calves or heifer calves, and it so happened that, during these troubled years, we were evidently having a run of heifers. We could only hope that the wastage from barreners would not be too great and that the heifers coming in, having been reared on the farm, would make good some of the losses. Some of them did, only to abort with their second calves. The occasional first calver also aborted.

Prior to this I had given considerable thought to what the plan for rearing heifers should be. Some of the old timers, who had survived the hard times and abortion storms of the 1930's, said the only thing to do was to run the heifers with the herd so that they would build up an immunity and become "pickled" to the disease. At least one very sound and practical vet also gave this opinion. My own vet, however, was convinced that the best thing to do was to keep the heifers completely away from the milking cows. The cows, he argued, would by now be carrying their calves because of the immunity they had built up and, the longer we could go without an abortion, the better the chance would be of the disease becoming attenuated.

There was, of course, no shortage of advice, most of it well-meant and most of it from people who thought they knew. It is most interesting to hear different vets discourse on whether or not a bull can and does transmit the disease, and to hear other people hold forth on the danger of the disease being spread by artificial inseminators. I believe I am right in saying that brucellosis in humans is very rare in inseminators. Throughout the years of trouble we used mostly our own bull, with some artificial inseminations. I never saw anything which led me to suspect either of these two methods of service as being in any way connected with our problem and, for what it was worth, the blood tests on our bulls only ever showed negative.

To be able to lay down a set plan of campaign in such conditions is difficult enough in itself. To keep to such a plan is well nigh impossible. As we lurched from crisis to crisis our thinking lurched with events as we were torn between the need to produce as much milk as possible and the need to get rid of the cows we thought would be the worst carriers. Having decided to get rid of the carriers it was no good then suddenly deciding to keep a particular animal when she aborted, just because she happened to be a very good one or from a valuable family.

By 1969, I am willing to admit, I had lost interest and hope and was utterly depressed about the whole miserable business. My wife, who is normally credited with being a restraining influence on my impetuosity and exuberance, now suggested that we should sue the Ministry. And I do believe that, had I shown the remotest inclination towards such an idea, she would have backed me all the way. Unfortunately, or perhaps fortunately, as the case may be, we were in no position to contemplate such action. It would have needed the backing of considerable financial resources which we didn't have. On the other hand, if we had been completely bankrupt it would not have mattered as we would have had nothing to lose. But we did still have something left, which we coulnd't afford to risk losing on such an impossible idea, which, in any case, could not really have been more than a gesture designed to spotlight the Ministry's appalling attitude and lack of understanding of the situation.

Our marriage is very much of a partnership in the best tradition of the great majority of farmers. We talk about what we are doing or propose to do, with the result that the eventual action of one is invariably with the full support of the other. Now, however, legal action being out of the question, my wife took the law into her own hands and did what she must have thought was the next best thing. She wrote to that nice Mr. Wilson. I thought nothing of the idea at all. But don't misunderstand me. I would have felt just the same way about it if she had been writing to that nice Mr. Heath. Politically, I am a member of that very powerful group known as the Floating Vote, and I am very pleased to say that I have never voted for the same lot twice running. After the way in which agriculture has been so shamefully betrayed by successive governments, certainly during my life-time, I cannot understand any farmer becoming an active worker for any political party.

In all fairness to that nice Mr. Wilson he evidently employed sufficient secretarial assistance for some junior assistant secretary to acknowledge the letter and send it on to the Ministry of Agriculture and Fisheries. There then followed an exchange of views between my wife and characters at the Ministry head-quarters in London who were, as I tried to convince my wife, being paid good money to send out these screeds of slush. It got nowhere, of course. Neither did what we were trying to do on the farm.

Having achieved no apparent progress by keeping the heifers strictly apart from the herd, I put it to my vet that there could be something to be said for the idea of "pickling" them. Whilst not being in favour of this he agreed that the problem had now assumed such awful proportions that he could understand my being desperate enough to try anything, and so, about 1968, we decided to let heifers have some access to the herd to build up an immunity.

I have already mentioned the run of heifer calves we had been having and, in the spring of 1970, we had twenty-two heifers out in the field. They had all either gone to bull or were about to go to bull, so that we looked for all twenty-two of them to calve and come into the herd between September 1970 and the spring of 1971. We had, of course, seen the herd once more go down in numbers as cows went off barren, or were slaughtered as part of the policy, and we reckoned on these twenty-two heifers bringing the herd back up to strength and forming the launching pad from which to surge forward bravely into the future and to meet the challenge of the 1970's.

All of these twenty-two heifers had had access to the herd to build up an immunity, and they all had their nice little ear tags to show that they had been vaccinated with S.19, and may the Lord have mercy on their souls. If heifers have souls that is. In the event, eight of them came into the herd and the last one aborted on January 8th, 1971. Enough was enough.

I suppose these heifers must have cost all of eighty pounds apiece to rear to that stage but to put the loss at fifteen hundred pounds is too easy. Who can put a price on the loss of production over the rest of what should have been their natural lifetime ? From time to time you will see sums worked out as to what contagious abortion is supposed to cost, but I am convinced that anyone who thinks he can work out such a sum must be either a financial genius or a fool. What it had cost us to that stage I really cannot guess. It is not that I am not prepared to work the figure out publicly in terms of increased borrowing. Few people outside of farming realise that many farmers have only been able to keep going of recent years because the increasing value of land has enabled them to raise more money on it. But this is not progress. It is hardly even survival. There are so many things to be taken into account that I do not think any man can assess what is the real price of abortion either to the individual or the nation. There is the cost of actual depreciation of capital, there is the loss of production, and the loss of potential. And God alone knows the cost in heartache and worry.

January 8th, 1971, was a Friday. I don't need to look at any old diary or calendar to confirm this because it is stamped all too clearly on my memory. For three nights I never closed my eyes. On the Monday at breakfast my wife said, "I suppose you wouldn't think of getting rid of the cows and turning the buildings into holiday accommodation ?" As casually as if she had asked me whether I wanted another cup of tea, I said yes, it was a good idea. We had been in the business before and knew what was involved, and we were in an ideal position, far enough from the sea to be away from the constancy of wasps and caravans and picnicers, and close enough to be there in five minutes. Long ago we had recognised that the holiday demand was going to be for self-contained self-service accommodation. In the day of the drip-dry, the deep-freeze and the oven-ready there is no need for life to be too much of a chore for the wife, and there is much to be said, particularly with a family, for being able to get up, eat and go to bed without the restrictions of a time-table. We had not done anything by way of adapting our farm buildings, so that, with their solid stone walls and slated roofs, they were just right for this purpose.

By the Tuesday the decision had been firmly taken, and on the Thursday, we sent what empty heifers we had to the mart to be sold or, more correctly, given away. As we loaded them in the lorry there were many memories that came flooding back. Memories of people connected with some of the better cows. Memories of loved ones who had gone ahead. And memories of all the brave hopes and ambitions which had once been ours. Shakespeare had a good deal to say about ambition and I reckon he knew a thing or two.

The idea was to go on producing what milk we could through the summer and be out of it by the autumn. The new plan was to have just over half the farm under corn, undersown with ryegrass to carry store lambs during the winter, and grazing cattle in the summer. There would, therefore, be no need to provide winter housing for any animals or to store winter fodder. Having that much corn would mean that there would be no place on the farm for sheep during the spring and summer and so, before the end of the month, we had, with real regret, sold our flock of Llanwenogs also.

I am prepared to admit that the next thing I did was criminal, yet there are some, I hope, who will not judge me too harshly. I am a founder member of the Pembrokeshire Records Society and serve on its Executive Committee. I never lose an opportunity to drive home the importance of the most ordinary old bill or receipt, notebook or letter. Even the ordinary record of the most humble countryman can be invaluable for the light it sheds for succeeding generations. When I was doing research for *The Sounds Between*, my book on the Pembrokeshire islands, I travelled hundreds of miles in search of people who might have old diaries or pictures of any sort, and I spent week after week searching through such records as were available at libraries, museums and the Public Record Office. All too often I found that something which would have been invaluable had been thrown away as being of no consequence. Yet, knowing and believing all this, such was my despair and misery, that I now set about destroying herd records with foolish abandon. Head-shrinkers could probably explain the attitude, but I cannot. The postal strike was on at the time and, as far as I was concerned, it was a period of great contentment which could have gone on a lot longer. Amongst other things I took the chance for a real clear-out of the place wherein I write. The kinder types have been known to say it is a place of some character—they don't say what sort of character. The others, less discerning, say it is just "a bloody jungle". As day succeeded day, and no incoming mail was added to the pile to be attended to sometime, the pile in fact began to dwindle until, it having been established that there still really was timber where we had memories of a desk having once been, the drawers and shelves came under assault. It was in this frame of mind, and with this opportunity, that I came upon so many things which I know should have been kept. Yet somehow I just wanted to put it all behind me. It was a closed book.

For a long time before this my wife had been telling me that I ought to have a blood-test for brucellosis myself. Looking back later I realised that I had been very ill for a long time and fighting against it without knowing against what I was fighting. Early in March I had been guest speaker at the annual dinner of the London Pembrokeshire Society. More often than not I go to London by train but, on this occasion, for some reason which now escapes me, I had gone by car. Admittedly it had been a convivial evening, but not nearly sufficiently so for it to account for the way I felt driving home the following day. It is a journey which, before the completion of the M.4, could be done comfortably in five and a quarter hours travelling time from door to door. Say six and a half hours including stops for a meal and petrol. On this occasion it took me nine hours and I felt like death all the way, almost too weary to change gear. Three or four times I just had to pull into the side of the road for a rest.

In the spring of 1964, shortly after we realised that our brucellosis troubles had started, my wife was ill. We had then been living here for four years and, when I 'phoned the doctor, I had to give him instructions as to how to get here. That's how often we called him out. My wife was really ill, sweating and with violent pains in the head. Our doctor at that time has since retired. He admitted to a loathing of all animal fats and declared that to drink milk was a filthy habit. He immediately asked whether we had had any abortion in the herd and I said, yes, just a little. It seems he suspected brucellosis. I didn't even know that humans could contract it and thought I had never heard such a foolish idea. To think that anything like that could ever come from our marvellous milk of which we had always been so proud. But he was such a good doctor, and had been such a real friend to us, that I humoured him over this little bit of nonsense which had somehow got into him. It turned out that my wife had sinusitis, and the blood-test for brucellosis proved to be negative. Which was only what any sensible person would expect.

Still, I knew now that humans, too, could suffer from brucellosis. I began to read and hear of an increasing number of cases. But it is the sort of thing which only happens to other people. Weary as I was, I never dreamed that I could have it. And I really was weary. I was trying to write a novel at the time. Always after supper I had been able to sit down and write until eleven o'clock or midnight, or even much later when the mood was on. And in that sort of mood I could also get up in the morning early and write. Now, I would settle down to write after supper, and about half-past eight, maybe having written a paragraph and maybe not, I would call it a day and go to bed saying I would get up early in the morning. Which, of course, I never did. It was as much as I could do to drag myself out of bed in time to take my son to meet the school bus and then, as often as not, I missed it. When I came back I would have my breakfast and sit down for an hour as though I had just done a hard day's work.

Everything had become a burden to me and writing, which I had always enjoyed so much, became the biggest burden of all. I had written three fifths of the novel, which is a long one based on the countryside, and had every incentive to get on with it, but it was now beyond me. When *Farm News* folded, in the spring of 1969, the *Farmers' Guardian* brought out a Welsh edition and I began writing features for them. As a farmer I liked the paper, and as a journalist, I enjoyed working for them. Yet, whenever I was doing features, it was one hard slog.

Journalistically I must be thankful. A number of agricultural papers have gone out of production in recent years, and other publications have cut down on their agricultural coverage, so that we have a position in which more and more erstwhile agricultural journalists have become freelances and are chasing less and less space. For my own part I have been fortunate enough to find all the writing work I want coming my way without having to go out and look for it. When I have finished my novel I have at least three books lined up for the first of which a contract is waiting. All of which is a fat lot of use if you don't have the strength to put pen to paper.

During the winter of 1970–'71 I had been having a course of injections for catarrh. When I went for the March boost I asked the doctor to take a blood-test for brucellosis. It was positive.

When I went to the specialist he put me through such inquisition as these characters usually do, only more so, and eventually came to that part of the catechism dealing with type of farming, herd numbers and so on. At this juncture I said, "Before you go any further, let me tell you now I claim to be the greatest living authority on the uselessness of S.19." To which he replied, "No indeed, you're not, I am !" It made me wonder how many other case histories he had in his files of farmers suffering from this disease who had been using S.19 in their herds and living in such a beautiful paradise for fools.

He said not to worry, he would cure me, and when I was better I would realise how ill I had been. He was right, too, but it took more than twelve months to do it.

Chapter 11

ON BEING A STATISTIC

Iᴛ is perhaps difficult for those who have enjoyed good health to accept that they can be ill. Apart from the serious, but short, illness of the spring in 1957, and the bit of trouble with my neck, which was not an illness anyway, I had always been on top of the world. My mother had been a nurse, and a very good one. Perhaps vaccination was not then what it is today, and I know she had seen some unfortunate results from it. Whatever the reason, she was firmly against it. I have never been vaccinated in my life and have never had even any of those infectious things which seem to come the way of all schoolboys. Whenever the school was closed because of mumps, measles, scarlet fever, chicken pox or anything else you can think of, I was the one with the bonus holiday. I eat my food disgracefully quickly, or so my wife tells me, but contrary to the worst predictions, and without wishing to rock the boat, I have not only never developed an ulcer, but have never even raised the wind. So maybe I am in no position to assess the degree of my illness.

Brucellosis, it seems, can take different people different ways. The most common symptom, apparently, is drenching sweats, but I never raised a sweat from first to last. The weariness was beyond description and the headaches almost unbearable. I would just have to go to bed of an afternoon, and six to eight aspirins wouldn't touch it. I tried all the recommended cures for headaches—"as advertised on T.V." which is the acme of recommendation for any commodity these days it seems—and all with the same fruitless results. There were, too, some aching joints, some days worse than others.

The treatment, I understand, is fairly standard. For three

weeks there is a daily injection of streptomycin, and for six weeks two tablets of tetracyclin four times a day. And I am assured that this is a very large dose indeed. The sort of dose which can and does cause side effects. And, in fact, did. I had finished with the streptomycin and was well on the way with tetracyclin when I went out to Skomer for a week with Sandy Copland. He had been having a pretty rough time and I reckoned that a spell of no telephone and no newspapers in the carefree breezes of this windswept island, with nothing for company but the seabirds and your own thoughts, would do him the world of good. It did, but it also played hell with my complexion. The antibiotics, it seemed, had completely desensitized my skin, and, by the time I came back to the mainland from this doctorless, shopless sanctuary, my face and hands were in a terrible mess. When that had been put right the next thing to do was to find a panama hat and sunglasses for the rest of the summer. Those who had never seen me before could never be quite sure whether I was the Aga Khan travelling incognito or just the Wizard of Oz. Those who knew me were merely amused.

Not the least of the drawbacks is the complete lack of sympathy. One author on the disease in *The Lancet* in 1970, said "The patient may look deceptively well, and psychoneurosis may wrongly be diagnosed, especially when the results of serological tests are negative or equivocal". It is a disconcerting experience, not to say somewhat embarassing, to be dragging about feeling like death's head at the feast whilst looking like an advert for the latest discovery in health food. I shall not fall into the trap of trying to dissertate knowledgeably on the medical terms and implications, but I must offer one or two thoughts. Long ago in my native Pembrokeshire, we had our own broad dialect. Many of the words and expressions still remain, and long may they continue to do so. One such was "feeling all to clush", which is about as descriptive and self-explanatory as any term could be. It meant, I suppose, all to pieces, all in a heap or all in a mess. If anyone sayd they felt "all to clush" you knew they felt rotten indeed. It wasn't the 'flu or anything in particular. It was just what it said, "all to clush". I have often used the expression myself and I wonder now how many of the old people used it when they were suffering from brucellosis without knowing anything about it.

There was, too, a complaint in the old days, known as "liver and hearty grow", in which the liver was understood to have grown fast of the heart. The more enlightened generations of the twentieth century have had many a good laugh at this piece of rustic nonsense. But brucellosis affects the liver, and I just find myself wondering sometimes whether in due course someone, more enlightened still, might not perhaps say that the old people were not so stupid after all.

Of course, on the question of lethargy, it is a disease which has been designed by Providence to accommodate the lead-swinger and the inherently lazy. But I am now beginning to wonder just how many folks in days of yore were classified in this particular category, when in fact they were suffering from a disease about which nothing has been known until comparatively recent times. Certainly I believe that, in rural areas at any rate, when a doctor cannot diagnose a patient's complaint, it should be standard practice to take a blood test for brucellosis. Admittedly this is not conclusive. The same *Lancet* article said "The value of these tests has been over emphasised". But, if they come up positive, that is at least a point from which to start.

The author points out that brucellosis is a "disabling and sometimes serious disease, and an important occupational hazard to farmers". That also includes their workers, of course, and, in particular, veterinary surgeons.

The infected discharge is mostly to be found when the cow calves or aborts and, to a lesser extent, when she subsequently comes into season. A vet, when attending such a cow during a difficult calving, is obviously at great risk, as indeed is anyone helping him. The cowman is at risk in disposing of the infected foetus and placenta, as the vet would be in removing a retained afterbirth, which is often necessary in cases of infection, and, of course, the cowman continues to be at risk whilst milking the infected animal. Her tail, particularly, collects the germs and these can be, and often are, flicked against the nose, mouth and eyes of whoever is milking her.

Another source of infection, which is not perhaps sufficiently appreciated, is the live calf from an infected cow. Children, whether from country or town, fondling such calves, as children will, are bound to be at risk. One farmer I know who had a bad dose of brucellosis, from which it took him eighteen months to recover, told me that he was convinced that he caught it from handling infected calves. He was driving calves from a collecting centre for a group of farmers every week.

Research work on the disease is in such comparative infancy that few figures are available as to numbers of cases and known sources of infection, and whatever figures are available are changing, almost literally, from day to day, as more cases and facts come to light. In any case, I cannot see that they are entirely relevant because there is a great danger in using statistics to try to prove anything. Certainly, in this case, it is hopelessly mis-leading for those who are trying to make light of the situation to use statistics to prove their point. If they have one that is. For the last reference I shall make to my own writings on the subject I would like to quote, in its entirety, my Ben Brock column of December 3rd, 1965. I wrote,

> "I've been thinking a bit lately about statistics, not forgetting what the gentleman said to the effect that there are lies, damned lies and statistics and this is a fact which everyone understands quite well.

Even so, it is still quite remarkable, knowing this, how some folks think that statistics are very important items indeed. I suppose they are in a way but the trouble starts when you find out that you're a statistic yourself. In fact I should imagine that to be a statistic is sometimes just about the worst thing that could happen to anybody.

The statistics already prove, I understand, that as far as planned re-deployment and shakeout are concerned, only so many of those affected have had to take on jobs at a lower wage and only so many are still unemployed with no immediate job in sight for them. So I hope these particular statistics can see their way clear to have a very merry Christmas on the strength of it.

I also hope that, when little four year old statistics are pulled lifeless from canals, their sorrowing parents will receive great comfort from the fact that they are the first statistics to be drowned at that spot during the last couple of years and not be too critical of purblind local authorities who have gone through life failing in their duty.

Any day now we can expect to see the statistics being produced by way of self-defence showing how many slag heaps have lasted for how long without moving, and how many children have passed through the schools beneath them during that time, which will be precious small consolation to anyone connected with the particular statistics at Aberfan.

I was also reading somewhere recently that the smaller the farm the more important it was to make more money per acre, so I have no doubt all the small farmers who come into this category of statistics will continue to be very hard-working statistics indeed and I wish them joy.

Now, of course, this sounds a very miserable column this week, and in fact, if you ask me, one of the worst things that could possibly happen to anybody in this life is for them to become a statistic, which is something I wouldn't wish onto anybody.

And I've got to thinking on these morbid lines because of what I was reading in this self-same paper last week about the milk producer-retailers taking things very much to heart about the health of the nation as far as it may or may not be affected by the drinking of untreated milk.

What's more, their concern is understandable because it says in the Book that ' where your treasure is, there will your heart be also '.

One gentleman, I see, is reported as having referred to the hullabaloo recently about the effects of brucellosis. Hardly, I fancy, an apt choice of epithet. As far as I've ever known, the word means ' a noisy confusion ' and, as far as I can see, until recently the noisy confusion has been on the part of those who fondly and foolishly believed that all would be well as long as farmers used S.19 for the vaccination of all their calves.

Certainly I can't think of any more confusion than there must be in the mind of whoever made the statement in that particular part of the same report, which says: ' More and more farmers are treating their stock against brucellosis.' How's that for a bit of confusion now ? Somebody tell me. Certainly whoever made such a statement couldn't have been reading the correspondence columns of this paper during recent months and this column in particular.

The gentleman, in fact, has quite a good case and there is a great deal in what he is saying that the risk is very slight and so much care is taken anyway.

Indeed I'm surprised that he doesn't look up the statistics to support his case. If they're not readily available and he would care to work it out for himself I happen to know of one poor miserable statistic for whom the bed had to be brought downstairs, and it's a pathetic story to hear how he was struck down by brucellosis in the full flower of his youth.

What I've never come across, however, are any statistics to show in how many herds using S.19 there's been a breakdown. And I think it's most surprising that the Ministry of Agriculture do not issue us with some statistics on this point because they would surely be most interesting.

Mind you, they wouldn't be one damn bit of consolation to the poor so-and-so's who had the misfortune to be the statistics in question. In fact, they're likely to turn out to be very angry and very unhappy statistics indeed. So maybe the Ministry figure that by not publishing the statistics on this score they will not be making people very angry and very unhappy.

Or maybe they just reckon that they wouldn't want to give too rude an awakening to those who are much happier with their heads buried in the sand.

Come to think of it I wouldn't like to say which is worse, for the head to be buried in the sand or to be just another statistic for whom nobody cares and whom nobody loves. Certainly to be a statistic must be a nasty and very unhappy experience".

Little did I realise that not all that long afterwards I was due to become a statistic myself. I should imagine that the number of farmer writers who are fully paid-up members of both the N.F.U. and the N.U.J. and who have experienced brucellosis both in their herds and by contracting it themselves must be extremely rare birds. In fact, if sufficient research could be done on the subject, I reckon I could be an absolutely singular statistic in my own right. But, I'm sorry to say, it brings me no joy whatsoever.

One thing for which I am thankful, is that I have recovered. I had a second course of treatment and am still troubled with headaches but, in between times, I have, like the farmer's cat in the old Pembrokeshire story, never been better.

Much more recently we have had reason to suspect that these continued headaches and debility in the family generally were being caused by lead poisoning from the water. Certainly replacing two hundred yards of very old lead piping has led to a spectacular improvement all round. So there's another statistic for anybody who compiles such things and it just so happens that the symptoms of lead poisoning can be very similar to the symptoms of brucellosis !

Whilst I was ill there were a few people who meant well (and they say the worst thing you can say about a man is that he means well), who tried to persuade me that I was better and that it was now in the mind and that I must get on top of it myself. I tried to persuade myself that this was right and to fight against it, but, now that I am better, I know that it was nothing of the sort.

My schoolboy son is a statistic, too. When it was found that I had brucellosis, my family, as a matter of routine, were called in for blood-tests. My son came up with a positive blood-test at the first time of asking. He had only mild symptoms and the verdict was that he should be all right as he had built up his own antibodies. Even so, he had to undergo more tests and have treatment because he had been, so typically apparently of the less serious cases, just "one degree under". It may not have been too serious in his case, but what build-up was this for a youth facing "A" levels in a fiercely competitive world ? How many young people have suffered a similar situation in the past, and indeed may be doing so now, without anybody knowing ? Do we still tell them they have growing pains when they talk of aching joints and limbs ? And the fact that, so it now seems, they can be cured does not entirely allay all the fears and soothe all the worries. Nor must it be forgotten that a cure is by no means certain and that the disease may flare up after injury or through further infection.

Chapter 12

DO PEOPLE MATTER ?

HOWEVER strongly a very small number of producer-retailers may feel about the hullabaloo, and however much the Ministry of Agriculture may dish out platitudinous hogwash about the progress they claim is being made by their pathetic approach to the problem of eradication, the fact is that it is now being established that there is far more brucellosis in humans than had ever been realised. The Ministry of Agriculture must accept the major share of the blame for this situation because, apart from the spread of the disease, which was encouraged by their first idiotic approach to eradication in the initial stages, the whole point is that the way to prevent the disease in humans is to get rid of it in cattle. To talk of pasteurisation of milk is only a small part of the problem. It still leaves the people at risk who are working with infected animals.

Not least of the tragedies is the fact that the entire national supply of milk, of which ninety-seven per cent is pasteurised, often comes under suspicion, and is brought into disrepute, by irresponsible outbursts, when only a very small fraction of the total supply is at fault. Nobody could have given better service to their fellowmen than have the milk producers of this country, mainly through the Milk Marketing Board, who have promoted the interest of the producer and served the very best interests of the consumer. If the consumer is in any doubt as to what he is getting for his money he might care to compare the price and food value of a pint of milk with the price and food value, if any, of a pint of pop. It is tragic that the hullabaloo caused by the indifference of the occasional producer-retailer should be allowed to jeopardize the good work which has been done over the years. Most assuredly no milk should be sold for human consumption unless it has been pasteurised, and the sooner this can be brought about the better, because brucellosis can never be cleared from the national herd the way the Ministry are messing about at the moment. The M.M.B. are doing what they can to help by taking routine samples of milk, and notifying the producer where brucella is found. This information, however, is confidential and, whilst being a guide, has nothing to do with eradication. All it really tells the farmer is that he has trouble which, in some cases, may not have been suspected. This milk, of course, which is sold wholesale from the farms, is all pasteurised and is safe as well as beneficial. The Board's testing scheme did not start in time to have alerted us to the infection which, all unsuspected, was in our own herd.

To advocate the pasteurisation of milk, however, is only a stop-gap remedy starting at the wrond end. It is no alternative to eradicating the disease in cattle. Perhaps, when the seriousness of the situation is brought to the attention of the public, something will be done to insist on action being taken. Unfortunately, there is a great deal still to be discovered about the disease in humans. The symptoms, as mentioned at the beginning, are Protean. As often as not they resemble the symptoms of other illnesses and the result is an incorrect diagnosis and unsuccessful treatment. An excellent article on brucellosis in the magazine, *World Medicine,* of September 1969, quoted a survey of fifty cases of brucellosis in children in the Macclesfield area. The fifty patients presented between them no fewer than forty-four different signs and symptoms. Only three of the cases were correctly dignosed by the general practitioners who referred them to the specialist, and thirty-two different diagnoses were suggested. But of fifty cases, "only seventeen were diagnosed in less than one month from the onset of symptoms ; in eight, diagnosis was not reached for more than a year ; and one poor child dragged round for five years before a name was put to her ailment. Many doctors now believe that, if you have a persistently depressed and tired-looking patient, you should at least think of brucellosis before sending him off to the psychiatrists or writing him off as a hypochondriac". (Maybe it would be no bad plan to check for lead poisoning whilst they are at it.)

The disease is reported as having affected people in many different ways with permanent effects being spoken of in some cases. A vet was reported in *Farmers Guardian* in April 1971, as having said that treatment in his own case had been totally ineffective, and claiming that one of his colleagues had been so badly affected that he had had to give up all his commitments, whilst two others had been crippled for life. One had been driven to attempt suicide.

Having referred to the danger inherent in drinking untreated milk it is only fair to emphasise that many of the people offering infected milk for human consumption are, not only completely in ignorance of the fact, but are indeed proud of what they have to offer. It is important to get this into its right perspective because literally thousands of visitors to the country every summer either buy milk at the farm-gate, stay in caravans on farms, or even have accommodation in farm-houses and drink untreated milk produced on the farm. Some of the farms, like some of the producer retailers, are brucellosis accredited. But this is not to say that the cows cannot have picked up infection since the last test and be transmitting brucella in the milk. Some farms may not be accredited and they may be one of those cases where infection has already happened but has not as yet made its greatest impact. Milk from such cows could be lethal. And it is perhaps permissible to wonder whether the person from the town could possibly be more susceptible than those who have more regular contact and the opportunity to build up a degree of immunity. The fact that the farmer or the producer-retailer is completely innocent is little consolation to the holidaying statistic who contracts the disease.

Things have improved, of course, from what they were. Back in the 1950's a young Carmarthen doctor in general practice, Dr. J. E. Davies, contracted the disease himself and it was this which prompted him to look more carefully into the cases of eleven of his patients who had repeatedly complained of ill-health and described symptoms similar to his own. It was established that the common denominator in some of the cases was milk coming from the same farm. But in those days, before the Milk and Dairies (General) Regulations of 1959 had come into being, nothing could be done to enforce either the removal from the herd of the infected animals, or the pasteurisation of the infected milk. Dr. Davies wrote on the subject in the *British Medical Journal* in 1957. He subsequently went to Canada and then to the U.S.A. There was no suggestion that the position was any worse in Carmarthen than anywhere else, but Dr. Davies thought that at that time there had been a decline in the drinking of pasteurised milk, due to the publicity given to tuberculin free milk, which had come to be regarded as "safe". No doubt any other doctor in general practice at that time would have found similar results as those brought to light by the pioneering work of Dr. Davies in his country practice.

Now, of course, as a result of these regulations, a pasteurisation order can be put on a producer-retailer when brucella is found in the milk. There are cases, however, where the regulations are not always enforced as they should be. Nor is the milk-ring tests on the milk by any means reliable. Furthermore, laboratories may fail to isolate Br. abortus from infected milk, especially when few attempts are made, and the cow may excrete the organism only intermittently.

Not long ago I knew of a case of a producer-retailer who could not go accredited because he was "having trouble," but his milk consistently passed the milk-ring test and could, therefore, be sold quite legally without being pasteurised. Naturally the man did not elect to advertise the matter from the housetops but, in the country, everything cannot be kept quite secret. The fact that any one or more cows could have been excreting brucella in between tests would have been of concern only to those who were drinking the milk. And I am not sure that many of them would have cared very much. It is staggering how many people adopt the attitude that they have drunk untreated milk all their lives and have never had any trouble, and so what. Yet the fact remains that until brucellosis can be eradicated from our herds, even if milk sold for human consumption is pasteurised, the families of every farmer and farmworker are at risk even if they never go near the cowshed or milking parlour. A man may live under a coal tip all his life and even an Aberfan in the next valley will not convince him that his own tip is likely to slip.

To come back to the case of the producer-retailer, however. Tests on the milk can sometimes take a long time to complete and the regulations provide that any legal proceedings must be commenced within six weeks of the date of sampling. By the time the law has taken its ponderous course, therefore, it is often too late to prosecute. Furthermore, because of the intermittent excretion, there is every chance that a re-test will prove negative.

Some local authorities even permit the sale of unpasteurised milk to rural schools in the mistaken belief that children could not contract brucellosis before the age of puberty, yet I know of a very bad case in a four-year old. There need be no apology from people for not knowing about something about which little is known, but it is a pity and very misleading when so many make statements which imply that they do know.

This, then, is something of the complexity and the magnitude of the problem with which we are now faced. A doctor has written, "Contagious abortion is hard to contain, and a vigorous educational programme aimed at the less-enlightened minority of British farmers is long overdue". As a farmer and a sufferer I have no quarrel with this statement, but I wish that he had also included the higher echelons of the Ministry of Agriculture. The Ministry churn out figures purporting to show how much it would cost to eradicate the disease. Over the years they have shown themselves to be so out of touch with the practicalities of the problem on the farm that they are obviously incapable of working out the figures. Their figures are, in fact, worthless, and little account seems to be taken of what it is costing the nation and the industry annually not to eradicate the disease, let alone the misery, heartache and distress.

Having been forced out of milk production myself it may be thought that I no longer have any interest. It is true, of course, that it cannot now mean to me what it did, but I like to think that I still care. I am feeling far better in myself and, where once I was completely dispirited and depressed, I am now once again full of the joy of living. I would like to think, too, that these pages have shown that I am not without a sense of humour. I have always tried to adopt the attitude in life that there are two things you can do, and that is either laugh or cry. And I found out long ago that if you cry nobody bothers anyway. Other people are not really concerned about our troubles, which is maybe where we go wrong. So, if there is no point in crying, then you might just as well laugh. There are some stupid people who mistake a sense of humour for flippancy. But to have a sense of humour is not the same as not caring or failing to take things seriously. What is more, I have certain fundamental Christian beliefs. And one of my beliefs is that, if Christianity isn't about people, then it isn't about anything at all. There is nothing incongruous to me about a sense of humour going hand-in-hand with Christianity. We do nothing for our cause by wearing black suits, looking miserable and condemning those who are not of our way of thinking. The big thing is that we should care, that we should be concerned, whether it be for the person who is waging the unequal fight against brucellosis in his herd, or whether it is the person who contracts the disease. The Christian believes that people matter, because Jesus said, "Another commandment I give unto you, that ye love one another". He rejoiced about just the one sinner that repented, and he cared about the one lost sheep. The individual meant everything to him. He was concerned about people, which is why he said, "Inasmuch as ye have done it unto one of the least of these my brethren ye have done it unto me". To him they were not just statistics. Can we really go on trying to delude ourselves that we are a Christian country if we don't care about people ?

There are those, of course, who will see this book merely as an exercise in self pity. In fact, nothing could be further from the truth. Having resigned myself to seeing our life's work in ruins the biggest impact, when the last of our cows (one of the Frances May family) had gone, was seeing the lights in other people's cowsheds at seven o'clock on a Saturday and Sunday evening and the best of British luck to them. What's gone is gone and we concern ourselves with the new venture on which we have embarked and adapting our farming to fit in with it. The only thing I have wanted to do is shed light on a problem as a result of my own experiences, and to warn others of pitfalls of the existence of which they are unaware.

Mark you, I have no doubt I shall be roundly condemned by many of the farming community because of some of the views I have expressed, but that is not the greatest of my worries. I know, and there are many others who know, that everything I have ever written has been aimed at, and in the hope of, achieving a better understanding of farmers' problems. Like the badger, of course, I can often be misunderstood, but I cannot keep quiet about something which is a real danger to farmers' interests, even though they may not be able to see it at the time.

Of one thing I am utterly certain, and that is that not many of those who disagree with me will be found in the ranks of those who, as I did, went through the fire and speak from bitter experience.

Is there any school in the world with such expensive fees as experience. ?

Who, indeed, can put a price on contagious abortion ?